HOW TO SPEAK SCIENCE

Gravity, relativity, and other
ideas that were crazy

until proven brilliant

Bruce Benamran

Translated by Stephanie Delozier Strobel

1 3 5 7 9 10 8 6 4 2

Virgin Books, an imprint of Ebury Publishing,
20 Vauxhall Bridge Road,
London SW1V 2SA

Virgin Books is part of the Penguin Random House group of companies
whose addresses can be found at global.penguinrandomhouse.com

Penguin
Random House
UK

Copyright © Hachette Livre (Marabout) 2016
Translation copyright © The Experiment, LLC 2018

Bruce Benamran has asserted his right to be identified as the author of this
Work in accordance with the Copyright, Designs and Patents Act 1988

First published in the United Kingdom by Virgin Books in 2018
First published in The United States by The Experiment LLC in 2018
Originally published in France as *Prenez le temps d'e-penser* by Hachette Livre in 2016

www.penguin.co.uk

A CIP catalogue record for this book is available from the British Library

ISBN 9780753548806

Printed and bound in Great Britain by Clays Ltd, Elcograf S.p.A.

Penguin Random House is committed to a sustainable future for
our business, our readers and our planet. This book is made
from Forest Stewardship Council® certified paper.

MIX
Paper from
responsible sources
FSC
www.fsc.org   FSC® C018179

To Jarod and Camille

CONTENTS

FOREWORD

by Michael Stevens

Hello! My name is Michael Stevens and I wrote this foreword . . .

. . . or did I?

How do you actually *know* that I, Michael Stevens, wrote the words you are reading right now? Is it because you trust the author and publisher of this book? Is it because you follow me on social media and saw me claim to have written this? Perhaps. But did you actually *see* me write this? Can you really be sure that you *know* anything at all? Isn't there always some possibility, no matter how remote or unbelievable, that what you think you know is wrong? For that matter, what the heck does it mean to *know* something in the first place?

Maybe we can start by agreeing that whenever you say you "know" something, you're simply saying that you believe that thing to be true—a description of reality as it is—for reasons that adequately convince you. That's pretty good. And it's that last bit, the reasons, that science concerns itself with. Science doesn't claim to deliver absolute philosophical truths, and it never promises to provide access to *all* answers, but as a means of investigation it has emerged as our most consistent, reliable, and adaptable weapon against mystery. As far as we know, it may also be the best we'll ever have.

To be sure, there are other ways to feel that you *know* something. For example, you could believe without question that a message in a dream or a voice in your head is true. Or your intuitive feeling that something must be true could be enough to satisfy you. But of all the methods we've got, the scientific method continues to be our most powerful. It doesn't shy away from being wrong or challenged—in fact, if an idea isn't willing and able to be refuted, it's probably not scientific at all!

And that's the beauty of science. It is not a tool for discovering truth; it's a method for reducing uncertainty. When you think scientifically, you ask, "How does the world work? How can I find out? And how can I become more and more certain that what I'm finding is ever nearer the truth?"

Science employs logic, skepticism, observation, and experimentation to not only find things out, but more importantly, to improve our confidence in (or heighten our suspicion of!) what we've found. It can often do such a good job that some ideas become so incredibly certain (like the roundness of the earth or the laws of motion) that it's tempting to think of them as verified truths. But that does a disservice to the spirit of the scientific method. No matter how certain or replicable or elegant a hypothesis might be, skepticism and doubt must always be maintained if we hope to continue improving our understanding of reality—not a skepticism that blindly denies everything, but rather one that is a constant force pressuring us to devise experiments and make observations to challenge our beliefs and to be ready to change them when better information comes along.

The book you are holding is a treasure trove of some of the wonderful things we have discovered using science and its methods. It was written by one of the greatest science communicators I've had the pleasure to watch and meet. Bruce isn't just good at explaining things; he also has a rare ability to turn explanations into invitations to learn more. His curiosity is contagious! I learned many new things while reading this book—but even more fun was the frequent experience of encountering a topic I thought I knew quite well, put into words that filled me with wonder all over again. I hope the same happens for you. Stay curious, and never stop wondering!

Michael Stevens is an educator, public speaker, comedian, entertainer, editor, and internet celebrity, best known for creating and hosting the popular educational YouTube channel, Vsauce.

INTRODUCTION

Be curious and take the time to get it.

Did you know that wearing shoes to bed increases your chances of waking up with a terrible headache? It's a rhetorical question—I am pretty sure you aren't aware of this phenomenon. It's a statistical fact that when people wear their shoes to bed, they are more likely to wake up with a pounding headache. Some alert readers quickly respond, "Dude! That's not the same thing." They are quite right.

If you say, "The population that sleeps with their shoes on often wakes up with a world-class headache," you are stating a piece of statistical data. Saying "if you sleep with your shoes on, you increase your chances of a painful awakening" is a logical statement. In the first sentence, something is noted. In the second sentence, something is predicted. If you confuse statistical data with a logic statement, you are confusing correlation with causality.

0. Correlation and Causality

Do you know where the most dangerous place in the world is? Just to be clear, let me establish right now that by *dangerous*, I mean "where you are most likely to die." That's probably a very reductive understanding of the word *dangerous*, but it's my book, so I'll do what I want. The most dangerous place in the world is nowhere other than in bed. Statistically, the place where you are most likely to take your final breath is quite simply a bed. Period. End of story.

I hope you agree that saddling beds with that kind of a reputation makes as much sense as saying "life is the world's deadliest sexually transmitted disease." It's playing with words, of course, but that's exactly the point of this chapter. Before I dive right in and give you your money's worth, let's take a few minutes—or maybe a few pages; I have no idea how quickly you read—to properly understand just how easily words can deceive us.

To say "beds are dangerous" is to say "the act of getting into a bed puts us in danger." While many people do die in bed, the beds are not the culprits. They aren't the *cause* of the danger.

Why do so many people die in bed? Many sick people and many elderly people spend more time in bed than do young, dynamic, thrill-seeking, recent college graduates. As a result, we can say, quite literally, deadly situations are more often encountered by people who are in bed, right where the Grim Reaper is lying in wait. There is definitely a correlation between the act of dying and the act of being in a bed.

Let's be clear: I am not challenging the principle of causality itself. It rains, the ground gets wet. The cause produces an effect. The rainwater falls to the ground and causes the ground to get wet. It's not very exciting, but it is definitely correct. A cause can even provoke a series of effects. In that case, we talk about a chain reaction: It rains, so the road is wet; the road is wet, so traffic moves slowly; traffic moves slowly, so I am late for work. Don't blame *me*. The rain started it!

In the same way, a cause can—and often does—lead to several distinct and separate effects. Going back to our first story: Those who hydrate themselves liberally and exclusively with alcoholic beverages during a festive soiree increase their chances of sleeping with their shoes on. At the same time, they also increase their chances of waking up with a headache, commonly termed a *killer hangover*. As a result, on a statistical level, it's quite true that the thing that caused such people to fall asleep wearing their sneakers or stilettos is the same thing that caused them to wake up full of vomit, introspection, and good resolutions: *"Never again. I swear, I will never touch another drop of alcohol again."* It is only because the two separate effects share the same cause. So you see, there is merely a correlation, not causality, between the shoes and the headache.

Confusing correlation with causality is such a classic logical error that you can say it in Latin, in not just one but two subtly different ways: *post hoc ergo propter hoc*,[1] and *cum hoc ergo propter hoc*.[2] Bonus!

Websites such as tylervigen.com list bizarre correlations between unrelated events using publicly available statistics. For example, when Nicolas Cage appears in films in the United States, there are more deaths by drowning; the amount of honey produced by bees rises when the number of arrests for marijuana possession falls. In the first case, there is a simple explanation for the correlation: Nicolas Cage plays in blockbuster movies that are often released in summer, which is when more people swim, and consequently, when more people drown. The business with the bees and marijuana, on the other hand, is probably a case of pure coincidence with no cause, effect, or explanation.

These examples are simple, so they are easy to figure out. However, reality is usually far more complex. It often conceals from us the connection between events. If I say, for example, high school students who regularly smoke marijuana often get bad grades in school, what can we conclude from this? Marijuana-smoking causes bad grades? Bad grades cause marijuana-smoking? Or how about this third scenario: Perhaps the student's home life is responsible for both the feel-good joints and the ugly grades? Hard to say—we usually prefer to simplify things and talk about vicious circles. In a circular system, causes can easily become the effects of their own effects.

In a nutshell, the whole problem is that reality is far more complicated than our theories. The smallest event may have many, infinitely small causes as well as multitudes of effects. This leads to the possibility of effects being much, much more significant than the causes—this is commonly called the *butterfly effect*.

Reality is too complex to be modeled perfectly. To get a handle on it, we first have to simplify things, extract some general rules, and then work out whether the rules are valid by way of reproducible observations and experiments. This is where scientists come in. When seemingly

[1]   After this, therefore because of this.
[2]   With this, therefore because of this.

legitimate, albeit lazy, assumptions and hasty conclusions become etched into the collective subconscious—young person from poor neighborhood = delinquent; video games = mindless trance; jock = idiot—we have to take a step back and separate what appears intuitively to be true from what is actually true. We need to be aware that our intuition is all too happy to play tricks on us at any given moment.

Scientists work in a special way. Their methodology is the main thing that distinguishes them from "believers": Scientists don't try to avoid data that suggest they are wrong. On the contrary, scientists try as hard as they can to expose their ideas to every possible counterargument. Frequently, it's not possible for scientists to prove they are right; they can show only that all the attempts to prove them wrong have failed. Science does not actually explain the reality of our world. Rather, science creates models that, under given conditions, tend to behave the way reality behaves. Sciences of any kind are only approximations. They offer only theoretical models, not reality.

1. Models and Reality

Imagine this experiment: You are in a laboratory and you would like to study the way a ball moves on a sloped surface. You have a small steel ball and a nice smooth, wide wooden plank. You elevate one edge of the plank using a block to make a 20° angle wrt (wrt = with respect to; that's how engineers say "in relation to") the floor. This experiment seems simple enough to manage. So simple, you could do it at home. But is it really all that straightforward?

First of all, let's consider the ball. It looks quite spherical, and it is indeed made of steel, but if you look at it through a powerful microscope, is it truly a perfect sphere? Or does it have surface imperfections, even if only thousandths of a millimeter high? Granted, it is made of steel, but how perfect is its composition? Could there be the tiniest impurity, even if only a few atoms at its core? And now, about the plank: Is it really possible for it to be perfectly smooth, right down to the molecular level?[3] It is sloped at a 20° angle wrt the floor, but can you be sure that it isn't

---

3   Does *smooth* even mean anything at this scale?

20.0000000001°? Regardless of the precision of the measuring instrument, there is always a threshold below which it is impossible to measure accurately. If you want to see for yourself, try using the handy-dandy disposable paper measuring tape from your favorite Swedish flat-pack furniture supplier to measure the thickness of a hair.

Next, let's consider the floor: Again, and for the same reasons, can it be perfectly horizontal? The air in the laboratory contains molecules: oxygen, nitrogen, carbon dioxide, etc. The air is in constant motion. The breath of the person conducting the experiment causes fluctuations in the air currents. The light that illuminates the entire laboratory, including the plank and ball, with its radiation also increases the temperature.

How can we say these details won't have any effect on the experiment we are about to perform? We just have to accept that the complexity of nature defies the most brilliant of minds and overwhelms supercomputers capable of incredibly complex calculations. Reality is, quite simply, beyond our comprehension. Accepting this concept is the first step in our journey toward understanding the world.

So, what do we do? Well, we have to simplify the experiment as much as possible. At the same time we confirm, through the experiment, that our simplifications and assumptions don't make much of a difference. Please don your safety glasses, a white lab coat, and grab a clipboard. Of course, you are also certainly welcome to do this experiment in jeans and a tank top. We'll assume clothing doesn't make a difference—unless you're wearing a magnetic jacket or something.

Let's assume, for example, the floor is flat, the plank smooth, and the ball a perfect sphere. If the laboratory is a reasonable size and the ball is sufficiently small, our assumptions won't affect the outcome. Check.

Next, let's assume the air doesn't interfere with the experiment. If we have "normal" air, not too much or too little pressure, at a reasonable temperature—commonly referred to as "standard conditions for temperature and pressure"—then the air does not, in fact, seem to influence the experiment. Check.

Finally, measuring the ball's movements means measuring its position. The ball, however, is three-dimensional, so measuring its position isn't necessarily a straightforward matter: Do we measure the position of the front of the ball? The middle of the ball? Let's assume that the ball is just a point—which has no size mathematically. Check.

Are you feeling the simplification yet?

OK, experimenter—see, you've already been promoted—you let the ball roll down the surface of the plank and measure its movement as a function of time, angle of the slope, and so on. Now, you can compare your results with results from equations developed by Isaac Newton. You'll quickly realize that Newton's calculations provide an excellent representation of the reality of the experiment—the observed values match the calculated values. Can you now say you and Newton understand the mechanisms that make the ball move? Yes . . . until proven otherwise. Try replacing the air in the room with a very dense gas, or replace the small ball with a gigantic ball, or replace the smooth plank with a very rough plank—or fly paper—and you will see that your assumptions and Newton's equations aren't suitable anymore. They won't come anywhere close to representing the new reality. No check.

Anyone working in science is constantly simplifying things in order to develop models. We call them *theoretical models*. These models must not—under any circumstances—be confused with reality.

A model tells the story about reality, in a given context, under specific conditions. Every model has its limits.

In this book, I'm not going to use mathematics as the common language to present a detailed description of all the various scientific models. This book is not a science course. The goal of this book is to present, as simply as possible, the different models that scientists have used in the past as well as those they use now. By using analogies, I hope to give you a feel for how reality behaves. This book also pays tribute to people who have devoted their life to gaining greater understanding of reality. Sometimes, it's just one small detail of reality. Sometimes, they never actually find the answers they were looking for. Sometimes, they find a bunch of new questions, and really, that can be just as exciting.

Caveat emptor—buyer beware.

MATTER

The true nature of reality eludes us.

Understanding reality, or at least seeking to understand it, begins with the question, What is it all made of? All that we see, touch, smell, hear—in short, all that we perceive. While this question may not be as old as the world itself (far from it), it *is* as old as people—the species, not the magazine. Even way back in antiquity (ancient time, before the Middle Ages), there were already two schools of thought that clashed on the subject of matter. On one side, we had Aristotelians; on the other side we had . . . intelligent people.

On the following page, we have a Focus Frame, a digression. You will find these throughout the book. Sometimes we'll need to go off on a rabbit trail, or a field trip, or some other wildly divergent path to discuss a wondrous detail. After the Focus Frame, we go right back to what we were reading about before the Focus Frame, but . . . all the wiser.

For a very long time, the Aristotelian school of thought was attached to the notion of *quintessence*—meaning "fifth essence." This school had decided that matter, whatever it was, was composed of air, water, earth, and fire[4] and the rest was just simply made of ether. Even though it's completely wrong, this theory had some merit in that you can follow the reasoning that helped explain physical phenomena such as gravity, combustion, and magnetism. Oh, yes. Aristotle was indeed familiar with magnetism. He didn't understand it *at all*, but he was definitely aware of it.

The other school of thought was led by Democritus—and of course, Leucippus before that. We know very little about this Leucippus. We don't even know if Leucippus was a woman or a man. Imagine a

---

[4]   The four elements come from Empedocles, according to Plato in the *Timaeus*.

Aristotle

For those of you who aren't familiar with my YouTube channel, e-penser, for several years now I've been having a major dispute with the most significant figure in Western thought—Aristotle. I don't question his greatness as a philosopher. I also allow myself to recognize his key contributions to a branch of mathematics called logic. However, for the record: I must insist that Aristotle was a really terrible scientist. For such a famous guy, he was hands down the worst scientist ever!

Some of you may say we should excuse him because of the era he lived in: He didn't know then what we know today, it's easy to judge an ancient sage from way back then, etc. Well, I say, "Um . . . no!"

Here is just a sampling of the idiocy we owe to Aristotle. Sadly, this pathetic rubbish was accepted until the end of the Middle Ages, even though simple common sense could have easily set things straight:

- Flies have four feet.
- The "fact" that women have fewer teeth—false—proves that women are inferior—also false.
- The gender of a goat is determined by the direction the wind is blowing when the goat is conceived.
- Eating hot food gets you a male baby (conversely, eating cold food gets you a female baby).
- A man's virility is inversely proportional to the size of his genitals.
- Inserting cedar oil, incense, and olive oil into the vagina is an effective female contraceptive—ladies, *don't!*

scenario over a snack of apple slices: "Auntie Leucippus, what great thoughts will you share with us today?"

The thought process began with a question that may seem rather obvious today, but it certainly wasn't at the time. Here is the question: If I take an apple and cut it into two pieces, then I cut one of those pieces in half, then I keep doing the same thing over and over; will I eventually get to the point where I have something that just can't be cut in half anymore? Would this elementary and unbreakable tidbit ultimately be the building block of matter?

Thus was born a scientific concept that jump-started revolution after revolution after revolution until . . . well, let's just say, we still aren't done with these "indivisibles"—in Greek, ἄτομος, or as we would write in English: *atoms*.

2. Atoms

I hope you don't mind if I state the obvious, but a person had to have huge cojones to suggest that an apple or a potato or a rock or a hair were all made of minuscule, indivisible components—it's just over the top! Furthermore, the differences in materials were determined by geometry, especially the later "hooked" atom theory—go look it up. This model used hooks that let atoms hold on to other atoms (reminds me of a gluon, but it's *not*—we're not there yet).

According to Democritus, there were these "atoms," yet their insanely small size prevented him from seeing them. What's more, there was the "void" between the atoms. Thus he was among the first people to conceive and construct a theory based not on observation but on logic alone. This earned him the pleasure of being detested by a certain Plato. So much so that Plato wanted all Democritus's works to be burned, so that nothing would remain for posterity. Amazing for a guy like Plato who didn't typically worry about whether his own ideas were correct or not. For a bit of context, while today no one would dare question the existence of atoms, I want you to remember that atomists were still fiercely debating with non-atomists until the dawn of the twentieth century—yeah, no kidding.

When the Library of Alexandria was destroyed—somewhere between the year 50 and the year 642, depending on which more or less hazy historical document you refer to—the Western world lost all trace of the great thinkers of ancient times until rediscovering them in 1453 when the Turks took over Constantinople. The fall of Constantinople led to European rediscovery of ancient texts and helped get the Renaissance rolling. During the Middle Ages, Aristotelian thought was the only school of thought supported by the Church. Thus it would be nearly two thousand years from Democritus to when Europe finally rejoined the epic saga of the atom. He who reopened the story was to be one of the last victims of the Inquisition: the famous and unfortunate Giordano Bruno.

Alchemy

It's a bit reductive to claim that during the Middle Ages, absolutely nothing went on about the atom. In reality, alchemists of the Middle Ages worked with twelfth-century Latin translations of Arabic writings. Without going as far as talking about atoms, alchemy was built on the idea of transforming the matter of one given pure element into another element (such as turning lead into gold). This is called *transmutation*.

Now about Giordano Bruno: This sixteenth-century Italian philosopher seemed to make it his life's goal to do everything possible to guarantee a warm welcome at the stake. He wasn't satisfied with merely rejecting the idea that the earth was the center of the universe. He wasn't happy with a heliocentric model with the sun at the center of the solar system. Well, actually, he did agree to put the sun at the center of the solar system, but he adamantly refused to place the sun at the center of the universe. Going further, he rejected the very concept of a center of the universe. He deemed the universe to be infinite. He thought each star was like our sun; they just looked smaller because of the distance. He also thought there were planets revolving around those stars, and he suspected those planets could harbor life. My dear reader, please understand that back in the 1500s, all these wonderful ideas were a recipe for one really *Big, Bad Barbeque*.

Bruno thought all matter was composed of indivisible elementary building blocks. He called them "monads."[5] In monads, we find the

---

[5]   Pronounced "MOE-nad." Like "Moe is mad," but with an *n* . . . and minus the *is*.

essential idea of the atom. (Bruno actually thought there were three fundamental types of monads: gods, souls, and atoms.) The monad is the physical counterpart of the *point* in mathematics, the basic unit in geometry. Bruno also believed that God was both the mystical minimum and maximum: "the monad, source of all numbers."[6] Clearly, for Bruno, the concept of the atom, aka monad, cannot be dissociated from philosophy and, therefore, religion.

The seventeenth century will see its great thinkers look up to the stars. For the most part, they will be "corpuscularists." Corpuscularists agreed with Galileo and Newton; they thought and accepted as a principle that matter is composed of minuscule indivisible bits. That's all well and good, but the true radical atomist of the day was Étienne de Clave, who along with Antoine de Villon and Jean Bitaud came up with a way to absolutely destroy Aristotle's ideas on matter. On August 23, 1624, they announced that on August 24 and 25, they would publicly support "four theses developed to refute Aristotle, Paracelsus, and the 'Cabalists.'" They made their statement using a series of posters that they hung all over the place. The first posters challenged anyone to refute their theses. Later posters stated the defense of their theses. Well, these three goofballs got themselves denounced as heretics by the Sorbonne. As an added bonus, the University of Paris would burn their work and anything that even came close to discussing atomism or Cartesian philosophy.[7] It just didn't end well.

People—the species, not the magazine—would have to wait until the eighteenth century for the idea of the atom to take its rightful place in history. Even though, twenty-five hundred years earlier, Anaxagoras —reputedly from Clazomenae—suggested this concept: Nothing could be totally destroyed; no thing just could come out of nothingness; only transformations were possible. At the time, it was "merely" a way of

---

[6] Giordano Bruno, *De triplici minimo* (*On the Triple Minimum*).

[7] Didier Kahn, "La condamnation des thèses d'Antoine de Villon et d'Étienne de Clave contre Aristote, Paracelse et les cabalistes" (The Condemnation of Antoine de Villon's and Etienne de Clave's Refutation of Aristotle, Paracelsus and the Cabalists), *Revue d'Histoire des Sciences* 55, no. 2 (2002): 143–98.

looking at the world, nothing more. It was a philosophy that was popular with a number of thinkers, particularly the Stoics.

However, in 1775, Antoine Laurent Lavoisier announced:

> For nothing is created, not in the operations of art nor nature, and one can state as principle, that for any operation, there is an equal quantity of matter before and after; there is an equal quantity of matter before and after the operation; that quality and quantity of the elements are the same, and there are only changes and modifications.[8]

OK, so we all know the famous saying, "Matter is neither created nor destroyed. It only changes form." You have to admit, that is kind of sexy. Antoine Lavoisier was definitely a great communicator, but Anaxagoras said it even better way back when: "Nothing is born or perishes, but already-existing things combine, then separate anew." Elegant.

The new thing, here, is that Lavoisier pretty much knows what he's talking about. He can clearly show the basis of his assertion by performing transformation experiments—primarily with gases. He demonstrates that the total mass of the components involved does not change, though the components themselves undergo radical changes in appearance. Lavoisier's work goes on to make him the father of modern chemistry. That's not really the point of this chapter. My key point is, by the end of the eighteenth century, they are finally asking questions about the nature of what matter is composed of—at its most *elemental* level, ahem—on the smallest scale they could imagine. All they needed was a color-blind fellow to lead them to the trail that would one day make Mr. Dmitri Mendeleev a planetary rock star.☮[9]

John Dalton was an Englishman, chemist, *and* physicist—all in one! In 1794, when he was twenty-eight, he realized that he was afflicted

---

[8]   Antoine Laurent Lavoisier, *Traité élémentaire de chimie* (*Elements of Chemistry*), p. 101, available at lavoisier.cnrs.fr.

[9]   When the peace sign ☮ appears, you are invited to serenely accept the idea and anticipate further discussion later in the book.

with a visual dysfunction that prevented him from seeing colors the way others do. So, he wrote a paper about it. This dysfunction was later called Daltonism. Besides that, he was deeply interested in what matter was made of. He was particularly interested in Lavoisier's work. In 1801, Dalton developed an atomic theory. According to him, matter was made of atoms; the atoms were absolutely unique for each element, and they could combine to yield structures as diverse as wood, water, an egg, air, and a potato. These atoms were indestructible; in this sense they were truly atoms, "indivisible" and immutable. They had always been and would always be. And, as Lavoisier stated, they can be recombined at will. While no one truly knows what inspired Dalton's theory or whether it was simply based on his own hypothesis, at least his theory had the good taste to explain quite a number of chemical phenomena, which is always a good thing for a developing theory to do. If a theory isn't based on anything in particular and doesn't serve any purpose . . . how do I put this . . . well . . . what's the point? *Cough, cough* . . . Aristotle.

Dalton's atomic theory explained Lavoisier's law of conservation of mass. If, in fact, the atoms themselves never actually changed but only recombined, then quite obviously nothing is created or destroyed, everything simply changes shape. His theory also explains the law of multiple proportions, which says that in a chemical reaction, the ratio between the proportions of the reagents and the products are whole numbers. Such as, running electricity through water (electrolysis) to break down two volumes of water—instead of using electrolysis to remove unwanted leg hair—yields two volumes of hydrogen gas ($H2$) and one volume of oxygen gas ($O2$).[10]

In 1803, Dalton sought to explain the differences in the behaviors of different gases. For example, water doesn't absorb nitrogen as well as it absorbs carbon dioxide.

Of course, the next thing Dalton wanted to do was calculate the masses of the elements. However, there was a bit of a hitch: It was absolutely impossible to make those measurements at the atomic level. Besides, what really interested him was the ability to *compare* the masses

[10] $2H_2O \quad 2H_2 + O_2$.

of the elements. So instead, he defined their weight relative to a reference element (giving hydrogen a mass of 1). This came to be called the atomic mass of the elements. Unfortunately, since he had no idea that gases such as oxygen and hydrogen existed as molecules of two atoms—dioxygen and dihydrogen—Dalton's calculations were really far off. So close and yet so far.

In 1811, Amedeo Avogadro helped solve this problem by offering this theory: If you have two different gases in two containers of the same size (volume) with the same temperature and pressure, the two different gases have the same number of particles in those containers. So, at the same temperature and pressure, when you have the same number of cute little gas particles, they take up the same space. No matter which gas it is! But . . . the masses are different. Aha! From this revelation, he deduced that some gases weren't single atoms; they must be combined atoms. That would make them, well . . . *molecules*. It might be good to remember that at that time, the atom was still only a theoretical concept. The relative masses could be measured even without the existence of the atom. A non-atomist school of thought called equivalentism emerged, with William Wollaston leading the movement. They attempted to construct a classification of masses, using oxygen as the reference and assigning it a mass of 100. Wollaston is gonna be taken down by the biggest rebel of all Russian scientists: Dmitri—*serious metal reverb here*—Mendeleev!

3. Good Old Science Dude: Dmitri Mendeleev

Dmitri Ivanovitch Mendeleev is absolutely the most *badass* of all the bearded Russian scientists of the nineteenth century—there must have been at least two or three. Born in Siberia in 1834, we don't know if he was the eleventh or fourteenth child in the family. What we do know, however, is that when Dmitri's father died, the family moved to St. Petersburg in 1849. Dmitri was fifteen years old at the time. His mother saw he had great interest and talent in the sciences, so she did everything in her power to get him into the University of Sciences. When he entered in 1850, at sixteen years old, he absolutely attacked chemis-try—for real! He was only *sixteen years old*. He went on to study under the direction of guys like Gustav Kirchhoff and Robert Bunsen—the fathers of spectroscopy.

In 1863, Mendeleev became a professor of chemistry. Six years later, he published a simple little[11] document titled "On the Relation-ship Between the Properties and the Atomic Weights of the Elements." Before going into more detail about this paper—a paper that ultimately revolutionized chemistry—let's take a little step back to understand what they understood about the elements *way back then*. We take a few steps back, then forward, then turn to the right, then to the left . . . now, you've got it! This book is a Macarena of science.

Spectroscopy

You've seen it on TV. The experts put a mysterious fiber in a machine, and the machine determines the molecular composition of said fiber. That's it. That's spectroscopy. Well, in Kirchhoff and Bunsen's day, it was a lot more difficult, but the principle was the same: project light at an object, break down the reflected light using a prism, and use the information to determine the composition of the object.

In 1817, the German chemist Johann Wolfgang Döbereiner drew attention to "triads." He classified elements with the same chemical properties according to their mass. He found when you take three (hence the name *triad*) consecutive elements, the mass of the middle element is the average of the masses of the two other elements. Whoa!!! This means, in a triad, the difference in mass between the one element and the next is the same. In 1859, the Frenchman Jean-Baptiste Dumas expanded this discovery to cover four elements, showing that "tetrads" also exist. He demonstrated that the difference in mass is constant between two consecutive elements with the same chemical properties. At this point, slowly but surely, people—the species, not the maga-zine—began to suspect there was some sort of *periodicity* (a repeating pattern) in the chemical properties of the elements.

[11]   Note the irony.

The next nail[12] in the non-atomist coffin is the telluric screw, aka telluric helix. This concept was developed by the Frenchman Alexandre-Émile Béguyer de Chancourtois. The telluric helix is a cylinder around which the elements are written on a diagonal, like the threads of a screw—hence the name—so the elements that have the same properties are displayed in a vertical line. The telluric screw showed a periodicity in the elements—that is, the difference in mass between similar, consecutive elements was constant. "Same difference"—a key concept—who knew?

Chancourtois's telluric screw confirmed and expanded on the tetrad concept from Dumas. Of course, the scientific community immediately

Hierarchy of the Sciences

Like it or not, there is a "hierarchy" in the sciences, an order by which one can rapidly determine the "importance" of a scientist. At the top, we have the completely theoretical mathematics and "hard" sciences: cosmology, astrophysics, particle physics, etc. Then come the applied life sciences and the rest: neuroscience, biology, chemistry. In the lower levels of this scientific pecking order we have geology and climatology. Then, way down deep, in the dark depths of the hierarchy of sciences, we find the human sciences—many don't consider them sciences at all—history, sociology, psychology, and so on.

Quite obviously, this hierarchy is a morass of prejudice. A science is a science. Period. End of discussion. While it may be possible to distinguish exact sciences (like mathematics) from inexact sciences (like psychology), in no instance should that make one science more or less important than another. A science can be recognized by its method: observations, hypotheses, theoretical models, experiments, validation. Who knows what amazing progress people—the species, not the magazine—lost out on because of all this prejudice? Discoveries made by women were often mocked and rejected. We'll talk about that later.

---

12   Nail—screw. Ha! OK . . . it's a screwy metaphor.

bowed down at the feet of such a genius. They erected statues to honor Chancourtois. They had parades in the street, so that people could . . . ah, no, just kidding. Truth be told, no one took his work seriously, because Chancourtois was a geologist and he used geochemical verbiage to discuss the elements. Their reaction was, How the hell can a geologist understand anything other than mud and dirt or rocks and caves? So Chancourtois was politely told to go back to his little cavern and let the adults handle the matter.

Fortunately, it wasn't long before a scientist "worthy of the name" took another look at Chancourtois's work. For the record, this scientist was a chemist, meaning he had some cred, so he could legitimately discuss elements without being completely written off. In 1864, the English chemist John Newlands published a periodic classification of the elements. He called it the law of octaves, and this is how he put it: The eighth element starting from a given one is a kind of repetition of the first, like the eight notes of an octave of music.

With the timing of this concept, Newlands actually beat out Mendeleev, and Newlands's name was recorded for all time as being the first person to correctly classify the elements and demonstrate their periodicity. So this time, the scientific community bowed down at his feet. They built statues. There were parades in the street . . . nope, just kidding. Newlands was either ignored or jeered by his peers in the scientific community. For one thing, his idea of the octave reminded them of the arbitrary ideas from antiquity—*egads!* Saying there is a periodicity every eight elements, the way there are eight notes in an octave . . . that's like saying there are five elements because five regular solids can fit into a sphere (i.e., the famous Platonic solids); really, the main reason it was not well accepted was because the rule worked for only the lighter elements up to calcium. Enter Mendeleev!

In 1869, Dmitri Mendeleev published "On the Relationship Between the Properties and the Atomic Weights of the Elements." This paper was a turning point not only for chemistry but also for people—the species, not the magazine—in their understanding of matter. The paper contains the periodic law for classifying elements. This classification is displayed in every chemistry classroom in the world: the periodic table of the elements—*more reverb here*. What's so great about it? A number of things, actually. The most amazing thing is, in 1869, no one was sure if atoms even existed. And no one, not even atomists, had any idea atoms might have *parts*! You know what else? Mendeleev was so far ahead of the pack that his hypotheses wouldn't be confirmed until the beginning of the twentieth century.

Mendeleev can be credited with seven strokes of true genius in his paper.

(1) *If you classify the elements according to their atomic weight*—well, they didn't use the words *atomic weight* at that time, but that's really what he was talking about—*you see a periodicity in their chemical properties*, such as stability and tendency to ignite or to attack metals.

(2) *If three elements have similar chemical properties*, you'll notice one of two things: *either they have similar atomic weights*, as in the case of iron, cobalt, and nickel, with atomic weights of 26, 27, and 28, respectively, *or the difference between their atomic weights will be the same*, as in the case of potassium, rubidium, and cesium, whose respective weights are 19, 37, and 55, with a difference of 18 each time. That's so cool!

The third point partially follows from the two preceding points: (3) *An element's position in the periodic table corresponds to its valence*. An element's position in the table reveals, at least to some extent, the element's chemical properties.

(4) *The most abundant elements in nature are those with the smallest atomic weights*. For once, you'd think Mendeleev would be content to look at his periodic table and say, The lightest elements are the most commonly available elements in nature. That's exactly what he *didn't* do. Instead, he proposed that this must be a law of nature, which has an explanation.

Based on our current understanding of stellar nucleosynthesis (see page 96), the cycle of "creating" atoms within the hearts of stars (isn't that romantic) explains perfectly why the lightest elements are the most abundant. You see, above a certain atomic number—call it the atomic mass, if you will—the atom breaks down (aka decays) rapidly into lighter atoms. For example, the existence of oganesson, atomic number 118, has recently been confirmed—and I mean *recently*: It was first confirmed in 2002. It wasn't even named until 2016. Please note that oganesson, atomic number 118, lasts only several *hundredths* of a *thousandth* of a *second*.

(5) *By simply knowing an element's atomic mass, you can know its chemical properties.* Mendeleev had so much confidence in his classification system, he claimed it could be used to describe an element even when only the atomic weight was known—that if someone wants to discover a new element and knows only the atomic weight, she knows what the element's chemical properties will be right then and there. A guy has to have the confidence of Muhammad Ali to say something like that and be ready to defend it.

Valence

A key chemical property of an element is its valence. Valence corresponds to the maximum number of atoms an element can form a bond with. So, for example, a hydrogen atom can bond with only one atom at a time, thus it has a valence of one. It is said to be *monovalent.* Our dear friend oxygen can bond with two atoms at the same time, so it has a valence of two. Oxygen is *divalent.* An atom that cannot form a bond with another atom is lonely—uh, wait, no—an atom such as neon that cannot form a bond with another atom has a valence of zero. The valence is associated with the fact that electrons are distributed around an atom. At the time, the late nineteenth century, no one imagined—not even in their wildest dreams—the existence of the electron. Just one more reason Dmitri Mendeleev was such an incredible *badass*!

Some atomic weights didn't fit properly in his classification system. Rather than question his system, Mendeleev concluded that (6) *some atomic weights were wrong and must be corrected.* Do you realize what he's saying here!?! "Whatever doesn't agree with my system is incorrect." *What!?* "You made the mistake! Fix it! If you have any concerns, simply let my system guide you to the correct answer." If Mendeleev had screwed up, his name would have been mud in Russian universities for decades! Worse, imagine if they'd had the internet! No worries though. Mendeleev was right.

Seventh and certainly not least (this point is definitely the one that shows the ultimate brilliance of Mendeleev's genius; it boggles the mind!), Dmitri noticed that there were gaps in some of the periods. So, he said that (7) *there are some elements that haven't been discovered yet.* You've got to understand what he's saying here! "I am so certain that my theory is right that the slightest disagreement with it can be explained only by your error or your ignorance." Again, he was right. The subsequent discoveries of gallium (1875), scandium (1879), and germanium (1886) matched Mendeleev's predictions remarkably well (at the time he called them ekaaluminum, ekaboron,and ekasilicon). So well, in fact, that it led the scientific community to accept and use the periodic law as a valid theoretical model. It only took them twenty years to catch on. Pfft!

A Day Late and a Dollar Short, Lothar

There's nothing to suggest Dmitri Mendeleev and Lothar Meyer ever knew each other. Lothar, a German chemist, also happened to be working on a periodic law for classifying elements. His first table, published in 1864, was expanded in 1868. It was not republished until 1870, a few months after Mendeleev's article. Soon, people were indeed talking about Meyer's periodic table for classifying the elements; unfortunately, it didn't go as far as predicting the existence of yet undiscovered elements.

A British chemist, William Odling, was also working on classifying the elements. His work was at least as advanced as Mendeleev's work. In fact, Odling arranged some elements, such as mercury and platinum, even better in his table. Odling also foresaw the discovery of more elements. However, Odling was a rival of Newlands and had been quick to scoff at Newlands's law of the octave. Because he'd been such a jerk, Odling's reputation was tarnished; other scientists wouldn't touch his work with a ten-foot pole. So, to this day, he remains primarily an illustrious nobody. It's sad, but haters gonna hate.

Well, that's all very nice and interesting, but what exactly makes Mendeleev a planetary rock star and not just some great Russian scientist? Here's the skinny: Like all rock stars, he had a crazy life story and an especially tumultuous love life. In 1876, even though he was married to a woman six years older than he with three children of their own, he fell in love with Anna Ivanovna Popova. Besides being the niece of his best friend, she was also much, much younger than he was—he was forty-two; she was sixteen.

In 1881, when she was twenty-one, he asked her to marry him. He swore he would kill himself if she refused—that's very romantic . . . in Shakespeare, maybe . . . but *not* in real life! That said, she must have thought him charming, with his hipster beard and all, because she married him the next year *while he was still married to his first wife*—as they say on the internet, "a normal day in Russia" (look it up on Reddit).

The divorce wasn't finalized until a month after they were married. Oh my! We could leave it at this: Mendeleev was a bigamist for a month. It was merely an administrative issue. He was already separated from his wife . . . yada, yada, yada. However, that doesn't take into account the Russian Orthodox Church. At the time, the Russian Orthodox Church had a great deal of influence. According to their rules, a person had to wait seven years after a divorce before remarriage. During that seven-year period, the divorced person was still considered married. Because of the bigamy issue, Mendeleev was barred from becoming a member of the Russian Academy of Sciences. Since divorce was rare at the time, Tsar Alexander II himself pulled some strings for Mendeleev. He was obviously greatly appreciative of Mendeleev. In the royal court, some complained the tsar was being too easy on Mendeleev. The tsar's answer? "Mendeleev might have two wives, but I have only one Mendeleev." #swagg

And thus, dear readers, the odyssey of the atom continues along its proper path until the atomist theory is completely "atomized"—groan—do you really expect my every attempt at humor to be successful? The atomist theory was eventually completely crushed, surprisingly enough by one of its major proponents, Joseph John Thomson.

4. The Electron

The history of the electron is intimately involved with the history of electricity—*electron*, *electricity*, they certainly sound related, don't they? The ancient Greeks noticed that when they rubbed amber with fur, the fur attracted tiny bits of stuff. Today, we talk about static electricity, but back then no one associated static cling with lightning (the only other electrical phenomenon "known" at the time). In 1600, the English physician and physicist William Gilbert would carefully examine magnetism in a book brilliantly titled *De magnete*. We'll talk about magnetism later; I only mention it now because Gilbert gave the name *electricity* to this habit of attracting tiny bits of stuff. He got the word from the Latin word *electricus*, which means "of amber." *Electricus* comes from the Greek word ἤλεκτρον, pronounced *electron*. It means "amber." That's why, in 1891, George Johnstone Stoney—what a name!—gave the name *electron* to the fundamental unit of electricity, the *atom* of electricity . . . or more correctly, *particle* of electricity—and . . . that's not quite right, either.

Now, before we go any further, just for kicks, let's go off on a rabbit trail that appears out of nowhere—let's talk about televisions! Our younger readers might be familiar only with flat screens that sit on their little pedestals. These wonderfully thin screens are only a few millimeters to a few centimeters deep from front to back. Our more senior readers remember, way back in olden times, when television screens and computer screens were almost as deep as they were wide. This presented significant challenges in creating a very large screen. In fact, in the 1980s, a forty-inch screen could easily be more than fifteen inches deep and weigh around ninety pounds. At that time, television screens didn't use plasma or liquid crystals; rather, they used a cathode ray tube (CRT), which was an improved version of an older invention, the Crookes tube.

William Crookes, a British chemist and physicist, fascinated by the way gases would or would not conduct electricity—whatever floats your boat—wanted to know what happened if he ran an electric current through a tube filled with gas at a very low pressure. He put a gas in a glass tube at low pressure. The tube had two electrodes (a cathode and

an anode), one at either end. When he tested it, he found that by lower-
ing the pressure of the gas and applying a high electric voltage, a ray of
light appeared to emanate from the cathode. He called it a cathode ray.
End of our rabbit trail from nowhere.

John Joseph Thomson was an English physicist who won the 1906
Nobel Prize for physics for, as the Nobel website puts it, "his theoretical
and experimental investigations on conduction of electricity by gases"—
aha! See, that rabbit trail wasn't a complete non sequitur. Thomson joined
Trinity College Cambridge briefly in 1876. Trinity College Cambridge—
that place is a veritable Nobel Prize factory! They have a total of thirty-two
laureates to date, including Thomson himself, *and* his son thirty-one years
later . . . yes . . . his son, too. Trinity College Cambridge, those hallowed
halls, where once roamed illustrious persons such as Francis Bacon, Lord
Byron, *James Bond*—he was an ornithologist, but still—Niels Bohr, and
even a certain Sir Isaac Newton. The Wikipedia list of famous students
from Trinity College is more than three pages long!

At the end of the nineteenth century, J. J. Thomson performed a
series of experiments on cathode rays. In 1897, he discovered it was
possible to change the direction of cathode rays by applying an electric
field around the tube. This provided definite experimental proof that
the ray was electrically charged. This also proved the ray was composed
of "particles," rather than waves as we will see on page 29—stay with me
here, you can sleep when we're done. Finally, after performing the
experiment using different gases, he concluded that the same charge
was pulled off the gas atoms regardless of the gas; the difference in the
charge was always the same. He had *merely* demonstrated, by experiment, the existence of the electron predicted by Stoney back in 1891,
though Thomson used the word *corpuscle*.

This discovery was a major turning point. It proved that the atomists had been right to keep fighting for centuries. At the same time, however, he proved the atomists were clearly incorrect. The discovery was
consistent with atomist hypotheses of the day. It confirmed their models. But remember, the atomists thought the universe was composed of
atoms, meaning insanely tiny, absolutely unbreakable, indivisible little

particles. The very word *atom* means "cannot be cut." The discovery of the electron demonstrated the electron was *smaller* than an atom. The electron is a component of the atom—rather ironic to confirm the existence of a subatomic particle first.

Now, don't be fooled. It's important at this point to remember: Atomist theories and hypotheses had permitted an understanding of matter for centuries, but their most basic assumption was false. It was time to update the theory. The atom was not the elementary building block of the universe. That fact doesn't negate that the atom—whether it's an elementary building block or a composite unit—has actual properties, particularly chemical properties, that were better understood. Existing theories of the day evolved to incorporate this new fact, rather than being completely dumped for some radically different theory. The easiest way to say it: Other than the fact that the atom actually is not indivisible, everything else we knew about the atom remained true. That's why they retained the term *atom*.

Because the atom was no longer the fundamental unit of matter, but some doohickey made up of other things, it was completely natural to start asking questions: What are these things? How are they arranged? That's exactly what Thomson asked himself the moment he made his discovery. Thomson took some particularly simple and elegant logical steps to envision what was going on in an atom. For one thing, he knew each atom contained a certain number of electrons. He also knew that electrons are negatively charged. Furthermore, he knew that an atom isn't electrically charged. Therefore, he concluded, the electrically negative electrons move more or less freely within a positive "soup," the whole atom being electrically neutral. If the electrons move "more or less" freely, it means two things: First, if you can emit a beam of electrons in a cathode ray, it's because electrons can be emitted—that is, they are not permanently locked into one location on an atom. Furthermore, electrons repel each other (because they have the same electric charge), but they can't spontaneously escape. That's because electrons are attracted by the positive soup (because it has the opposite electric charge).

Thomson's model came to be named—but not by him—the "plum pudding" atomic model. In this model, the electrons are part of an atom like dried plums (aka prunes) in a pudding: not completely captive, yet not totally free—like the guy down the street with the ankle bracelet. There are other similar names for this model, such as the blueberry muffin model and the chocolate chip cookie model. It certainly shows that when purely scientific language is left behind, local culture comes into play. As far as Thomson was concerned, he titled the article: "On the Structure of the Atom: An Investigation of the Stability and Periods of Oscillation of a Number of Corpuscles Arranged at Equal Intervals Around the Circumference of a Circle; with Application of the Results to the Theory of Atomic Structure." You've got to admit, "plum pudding" sounds a lot sexier.

We should note that the Japanese physicist Hantaro Nagaoka completely rejected Thomson's model, arguing that opposite charges were impenetrable—that is, charged particles cannot be situated within another or move through one another or touch each other. In 1904, he proposed a model—largely unnoticed at the time—in which the electrons revolved around a positive charge like Saturn's rings. For his model to be correct, essentially all the mass in the atom would have to be located at its center.

Several variations of the model were developed. Some talked about a positively charged cloud rather than a soup, whereas others imagined that the electrons run in circles inside the atom, but Thomson's model remained relatively unchanged until it was completely invalidated, in 1909, by an experiment directed by one of Thomson's own former students, Ernest Rutherford, and performed by Hans Geiger and Ernest Marsden. The experiment has come to be known as the Rutherford gold foil experiment, or the Geiger-Marsden experiment. This experiment resulted in the discovery of another important part of the atom: its nucleus.

5. The Nucleus of the Atom

In 1907, Ernest Rutherford, a former student at Trinity College—yes, another Trinity alum—worked with Hans Geiger in Manchester,

England. Together they invented a counter that detects α (alpha) particles, emitted by radioactive materials—yes, indeed, I'm talking about the predecessor of the Geiger counter. Rutherford was enormously interested in radioactivity—as was Marie Curie, of course, but she deserves a whole book just for her. He was convinced that α particles were actually helium atoms that had lost their electrons. He demonstrated it in 1908. Here's a quick summary: He isolated radioactive material in such a way that only α-particles would be collected; he then analyzed the spectrum of the gas he collected from around his device; from this, he determined it was indeed primarily helium, meaning the α-particles had regained their neutral charge—that is, they had regained their electrons, and so had "become" helium atoms. A little story about Rutherford: That very year, 1908, he won the Nobel Prize for chemistry "for his investigations into the disintegration of the elements, and the chemistry of radioactive substances." Rutherford was happy enough about it, but he was disappointed that he hadn't received the Nobel Prize for physics, because in his opinion "all science is either physics or stamp collecting." See what I mean when I say some people have a hierarchy for the sciences?

In 1909, Rutherford decided to aim a beam of α-particles at an extremely thin piece of gold foil—six millionths of a meter. Compared to a meter, it's comparing an eyelash to an Olympic swimming pool. Actually, he already had performed the experiment using a thin sheet of mica, but he performed it again in a more precise manner with Geiger and Marsden using the gold foil. According to Thomson's positive soup model, nothing should hinder α-particles from going in a straight line right through the gold foil. Rutherford put gold foil in front of a screen coated with zinc sulfide. Zinc sulfide makes a small flash of light, aka scintillation, when α-particles hit it. Imagine the surprise the men got when, after a few minutes, instead of the nice sharp mark they expected to see on the screen behind the foil, they saw scintillating points of impact spread all over the place, even at large angles—up to 90°—with respect to the original beam! Something was definitely deflecting the α-particles! I wrote "Imagine the surprise," but Rutherford himself expresses it best: "It was almost as incredible as if you fired a fifteen-inch shell at a piece of tissue paper and it came back and hit you!"

The analogy is perfect. Nothing should have prevented the α-particles from going through the gold foil, let alone deflect them. Rutherford had to think about it for two years before suggesting an interpretation of this phenomenon. Since 99.99 percent of the particles had *not* been deflected, his conclusion was that matter was primarily composed of *void*; that was the first crack in the bowl of positive soup. He also concluded that the remaining 0.01 percent of particles deflected were repelled by positive charges. Therefore, the positive charge of an atom was concentrated in a very small portion of the atom's volume, one hundred thousand times smaller than an atom. Rutherford concluded that all of the positive charge in an atom, as well as practically all of its mass, was located in an extremely small nucleus in the middle, surrounded by void, with the electrons gravitating around the nucleus and the electrons establishing the "perceivable" size of the atom. Thus he recognized characteristics very similar to those proposed by Nagaoka seven years earlier and presented a new atomic model, the Rutherford model—also known as the Rutherford-Perrin model, or best known as the planetary model.

In this model, the nucleus of the atom is like a sun with electrons gravitating around it like planets. At this stage in scientific research, an electron was represented as this tiny little thingy, a rigid sphere to be exact, without going so far as asking what the little sphere was made of. Here's the difference between the solar system and an atom—besides size, come on, give me a break: An atom isn't flat; electrons don't travel in a flat plane around the nucleus. Let's say it's more spherical and the electrons are located at a distance from the nucleus, revolving very quickly all around the nucleus. Thus the electrons define the shape of the atom.

For Rutherford, therefore, the electrons orbited in a circular manner through the void all around a nucleus. This posed a number of problems. First of all, if you think about the way a planet revolves around a star, you can imagine an electron revolving around a nucleus, but because of "Coulomb's law"—opposite charges attract—for such movement to exist, the electrons must be continually accelerating. Why? The act of revolving

The Sphere

Since sooooo many things resemble spheres or balls or are considered to have the shape of a ball or a sphere, maybe it's useful at this point to talk about the particulars of a sphere.

First, imagine a goat—a *what*?! Yes, a goat, in a garden, attached to a stake with a rope one meter long. Your goat can walk around in all directions up to one meter from the stake. If she pulls the rope to its full length and walks around munching only the grass beneath her nose, she will leave a mark in the shape of a circle with a one-meter radius, with its center at the stake. Simple enough. Now, if she snarfs down all the grass she can reach, from the stake all the way out to the very end of the rope, she will mark a disk on the ground—same center and same radius as the circle. So far, your goat can only move around in two directions, or two dimensions: left to right, front to back. What if she could move in a third dimension, meaning up and down—wouldn't that be fantastic? A levitating goat!? Now that she can fly, if she pulls to the end of her rope in three dimensions, her nose will trace a sphere instead of a circle. If she gorges in 3D on everything from the stake to the end of the rope, she eats a ball rather than a disk. Officially, a sphere consists of all the points in space at the same distance from a given point called the center, like the surface of a bubble. A ball is *all* the points between the center of a sphere and the sphere itself, like a foam ball.

So maybe you haven't learned anything particularly profound in this Focus Frame, but it may help you understand the shape of atoms, planes, stars—and even soccer balls, which aren't actually balls; soccer balls are actually air-filled spheres. Remember that.

around the nucleus costs the electron energy. If an electron doesn't accelerate, it will get closer and closer to the nucleus and slow down enough to crash into the nucleus. If that's the case, then atoms would be unstable! We wouldn't be here to talk about them—by *talk*, I mean "write," as far as I'm concerned, and "read" as far as you're concerned. Therefore, electrons must accelerate. Here's another hiccup: James Clerk Maxwell, the "ultimate boss" of electromagnetism, showed that an accelerated particle radiates energy, meaning it loses energy. Therefore, it will slow down, eventually running into the nucleus, and we all crash and burn again! Thank goodness, that's not the case. My point is, the model stinks. Fortunately, Rutherford was poised at the brink of making two more discoveries that would, once and for all, place his name in the pantheon of guys who found out stuff.

Jean Perrin

Did you maybe half-notice that the Rutherford model is also called the Rutherford-Perrin model?

Allow me to take a few lines here to pay homage to Jean Baptiste Perrin, unknown to all, or nearly all anyway. Jean Baptiste was a French physicist, chemist, and politician who discovered the electron in 1895 when he was twenty-five years old—two years before Thomson, *and* Thomson was forty-one years old when he discovered the electron. Perrin discovered the electron while studying how cathode rays are deflected by an electric field, just like Thomson. So why did Thomson get the credit? Simply because Perrin wasn't seeking to understand the negatively charged particle; he was trying to prove its existence. Perrin's goal was different from Thomson's. Perrin was trying to resolve a debate in the scientific community regarding whether the cathode ray was made of particles or waves. By showing there was a negative charge, he demonstrated that the cathode ray was made of particles, because waves don't carry an electric charge.

Thomson built on Perrin's work to identify and characterize—fancy word for "describe"—the electron.

Later, in 1908, Perrin was the first one to confirm Einstein's work on—we'll get there—Brownian motion, thus confirming the existence of the atom. Later, in 1919, he would correctly predict how nuclear reactions provide the source energy to stars. Not bad for a little guy from a modest family in Lille. Here's *how* modest: His mother's family didn't have a lot of money; in order for her to able to marry Perrin's father, an artillery captain, she first had to obtain special permission from Napoleon III himself. Social rank was certainly no laughing matter at that time.

As we know, Rutherford just loved to shoot α-particles all over the place. While he was thus engaged, he noticed when he bombarded hydrogen with α-particles, a what-cha-ma-jiggy with a positive charge was released. He also noticed that a lot more of these what-cha-ma-jiggies were released when he was strafing nitrogen with α-particles. This led him to conclude that there were hydrogen nuclei in the nitrogen nuclei. Hydrogen is the atom with the lightest nucleus, so he decided to name the nucleus πρῶτον, *proton*, which is the Greek word for "first."

No big deal, Rutherford had *only* just made two foundational discoveries. First of all, he had discovered the proton, the positively charged particle found in the nucleus of the atom. In addition, he had discovered that the atomic nucleus is not homogeneous; it is a composite. This discovery creates an enormous problem in understanding the fabric of reality—in my opinion, one of the greatest things about science is that every new answer brings up new questions!

OK, let's say the hydrogen nucleus is made of a positively charged what-cha-ma-jiggy called a *proton*. Let's also say that any atoms more complex than a hydrogen atom are made of a composite of multiple hydrogen atoms—thus multiple protons—stuck together to make a relatively heavy "supernucleus." At that point, the burning question for Rutherford became: How the hell can two protons be stuck together!? How is there any kind of stability when they both have a positive charge and should be pushing away from each other hard as they can? And you better believe it, on that scale, they repel in a *big* way!

In 1920, during a conference on the structure of atomic nuclei, Rutherford dreamed up a substructure composed of a proton and an electron joined to each other to form a neutral structure that was very compact, with the proton and electron charges in close proximity—would Nagaoka approve? This structure would counteract electrostatic repulsion between protons and hold them within the nucleus. Rutherford was thinking of some kind of composite particle: essentially the same mass as a proton, because an electron is very light compared to a proton, and about the same size as a proton, because an electron is absolutely, ridiculously small, compared to a proton—that was before they

Coulomb's Law

I mentioned Coulomb's law on page 27. At this point, it's probably a good idea to share with you just how big of a problem this is going to be—I wanted to write *enormous* with twelve O's to indicate the enormity of the problem, but my spell checker didn't like it, and I didn't want to disable it, just in case I made any mistakes later on. Anyway—there were already two laws from Coulomb, one for electrostatics and one for mechanics. How many laws does one guy need, anyway? Well, the Coulomb's law we're talking about here is the one for electrostatics. The law is named after the French physicist Charles-Augustin Coulomb, who stated the law in 1785—more Macarena of science: step back in time from Rutherford, now to the side, hey, Macarena! By the way, do you remember Stoney? He's the one who predicted the particle of electricity. He also suggested using the electron's charge as the unit for measuring electrical charge. Ultimately, the scientific community preferred to use something else for the unit for electrical charge—"the quantity of charge transferred in one second (1 s) across a conductor in which there is a constant current of one ampere (1 A)," as the dictionary has it. They called it a Coulomb (1 C).[13] Coincidence? I don't think so.

Among other things—and why he comes up in our dance now—Coulomb established that electrical charges repel and attract like north and south poles on magnets. Electric charges can be negative or positive. He determined that they attract each other when they are opposite signs and repel each other when they are the same signs. Of course, Coulomb wasn't content with just saying that; he also provided the formula for calculating the force of this attraction-repulsion. And, as we shall soon see, this formula is incredibly—or not—similar to the equation that describes the law of gravitation.

[13]   The definition is written: $1\,C = 1\,A \cdot 1\,s$, which is beautiful.

figured out that the electron had no size at all, but I'm getting ahead of myself. Bottom line, he was thinking of some kind of neutral gadoodle, a lot like a proton, but neutral. It was well thought out. Ultimately, it was completely wrong, but it was well thought out. It was time for the next bright young talents to dive into a popular new theory—quantum mechanics. They will find that idea of an electron in the nucleus didn't make any sense, and it was incompatible with other promising concepts, like the Heisenberg uncertainty principle. Rutherford's idea was completely wrong, but it was a good effort. Rutherford had already done enough for science. Enter Chadwick.

What About the Bohr Model?

Some folks are probably surprised that I haven't yet discussed the Bohr model. Indeed, Neils Bohr greatly improved on Rutherford's atomic model a few short months after its creation. That means he deserves a place in this chapter. I hesitated because it's tough to talk about Bohr's model without talking about Max Planck's work, and it's hard to talk about Bohr's work without actually talking about Bohr, so please bolster your patience and we'll return to the subject when we talk about quantum mechanics.

James Chadwick was one of Rutherford's most brilliant students—at Cambridge University, though, not at Trinity. Chadwick was at Cambridge University in 1920 when Rutherford was searching in vain for an atom-ish thingy with no electrical charge and a nuclear mass of 1 that could hold together the nuclei of more complex atoms. Chadwick listened carefully and kept the idea in the back of his mind for *years*, until one day in 1932, when he couldn't explain the results of one of his experiments. Until then, scientists had made do with an atomic model with electrons, protons, and "nuclear electrons."

In 1932, James Chadwick performed yet another experiment with α-particles. But wait! Before that—more Macarena—in Germany in 1930, Bothe and Becker bombarded lithium, beryllium, and boron with α-particles. They found that when bombarded, these three elements released highly penetrating radiation that Bothe and Becker thought were gamma (γ) rays—yes, of course, later, we'll talk more about Marie Curie and radioactivity, and you're right, γ-rays created the Incredible Hulk. Curiously, the mysterious rays emitted from lithium, beryllium, and boron seemed to have a lot more energy than usually seen in radiation. Also in 1930, Irène and Frédéric Joliot-Curie—yes, the daughter and son-in-law of Marie Curie; you'll get a complete rundown on radioactivity but not right now—tried to solve the mystery of the famous γ-rays.

Ultimately, in 1932, it was Chadwick who determined they were absolutely *not* γ-rays, and he had tests to back it up. He precisely measured the energy of the radiation and found they were particles with the same mass as a proton but with a neutral electrical charge. Chadwick remembered Rutherford's suggestion and realized it wasn't some sort of proton-electron composite, but another elementary particle! Like the proton, only electrically neutral. Naturally, he named it a *neutron*—we now know that neither the proton nor the neutron are elementary particles, but you've got to admit, at the time, it was really first-class work. For the discovery of the neutron, he received the Nobel Prize in physics in 1935. It's interesting to note that Chadwick originally wanted to be a mathematician, but he got into the wrong line to register for classes—yes, lines, you know, like at the department store, where people stand one in front of the other; you didn't register online back then. When he realized his error, he was too embarrassed to admit his mistake, so he decided to stay where he was and study physics. Allow me to introduce James Chadwick, card-carrying member of the Major-Balls Club!

Well, we're getting there bit by bit. At this point, we have atoms that have a nucleus, in which we find protons and neutrons, to form the nucleon family. We also have electrons "revolving" around the nucleus, far away from the nucleus. Essentially, all of the weight is found in the nucleus—the weight of a neutron is the same as the weight of a proton. We find that atoms have equal quantities of protons and electrons, charged positively and negatively, respectively; thus the atom, as a whole, is neutral. Elements are identified by the number of protons in the atom's nucleus—one for hydrogen, two for helium, etc.—and while the number of neutrons may vary, it doesn't change the kind of an atom it is—having one or two neutrons in a hydrogen nucleus doesn't change the fact that it is a hydrogen atom, but it does change some of its properties. When there are extra neutrons, we talk about isotopes of an element. You can take a breather here; we're finished—for the moment—with atoms. On to a lighter subject! And what can be lighter than *light*?

While telling you about the fabulous era in this section, which was so rich with discoveries, I glossed over a long-standing and very heated debate about the occurrence of α-particle radiation, which you heard about earlier; however, the issue relates more directly to questions about light: Is it a flux of particles that one can model as tiny particles traveling at very high speeds? Or does light propagate like a wave, just like the waves you get when you push Bob into the deep end of the pool to see if he can swim?

LIGHT

When we think about light, we come to the realization
that we don't understand anything at all.

The desire to understand the nature of light is as old as the desire to
understand the nature of matter. Since ancient times, Plato and Aristotle's
epoch, we've been aware of many of light's characteristics. For instance,
light propagates in a straight line in a homogeneous medium, and the
law of reflection explains how you can see your own face reflected on the
surface of water. One hundred years after Plato and Aristotle—at around
300 BCE—Euclid produced a work titled *Optics*. In *Optics*, he described
how light works and pronounced his geometric laws for light.

For Plato, Aristotle, and Euclid light was nothing more than a tool
for seeing; essentially, just one of the senses. They thought that rays of
light came out of their eyes and were intercepted by the objects that
were illuminated. This idea would persist pretty much until the Renaissance. Apparently, no one wondered why they couldn't see at night—all
righty, then.

However, we *can* see that in the second century, Claude Ptolemy—to
whom we own the geocentric model of our solar system—which, by the
way, wasn't much of a "solar" system at all, since he thought the earth was
the center—studied reflection and refraction of light. He determined that
the angle of refraction depends on the density of the medium. Ptolemy's
mathematical formulas on light were used for nearly fifteen hundred years.

As was often the case in what we now call the Western world—that
is, Christian Europe of the Middle Ages—we had to wait until the
Renaissance to make any more progress in the area of optics and understanding the nature of light, ahem. However, in the Orient, by the end
of the tenth century, enormous advances had been made, mainly by Ibn

Reflection (Diffuse or Specular), Refraction

Go stand in front of a mirror. You are lit by a lamp, you are facing the mirror, and you see yourself in the mirror. Bravo. What's really happening? We'll unravel a little more later, but right away, we can state that rays of light are *not* coming out of our eyes, contrary to what Aristotle thought. We understand that the lamp is the light source. The light leaves the lamp, comes to you, and hits your surface (your skin), the surface of your clothes, the floor, the furniture, etc. Part of the light is absorbed by your skin and part is sent back, or reflected. That's why you can see yourself: It's just as if your skin itself was emitting light. This bouncing of light is called *reflection*. Part of the reflected light that bounces off you goes and hits the mirror. At the mirror, the light is reflected again, but in a "cleaner" way. Then, some of the rereflected rays arrive at your eyes, and that's what you see.

When I say a "cleaner" reflection, I mean the rays of light that come back to you maintain the axes of symmetry; this allows reflections in a mirror to look like the lighted object. But if a mirror reflects and skin reflects, why doesn't skin look like a mirror? Why can't I see myself reflected in someone else's skin? Let's just say for now that the light reflected by the skin is *diffuse*; the rays go off in all different directions. The rays reflected from the mirror are reflected in a much more organized and homogeneous manner, called *specular reflection*. We'll cover that in more depth later.

As for refraction, it's a little harder to explain, but it's fairly easy to observe. You've already noticed it when you put a straw into a glass of water. The straw appears to be broken where it hits the water. That has to do with the fact that the light bends when it goes from one *medium* to another—such as when it passes from vacuum into air or from air into water.

al-Haytham. His Westernized name is Alhazen; his full name is Abu Ali al-Hasan ibn al-Hasan ibn al-Haytham. I'm going to shorten that to "Ali," no offense intended by using the nickname. Around the year 1000, Ali generated a work of not less than seven volumes called the *Book of Optics*. I tell you, Ali was no slacker . . . not in any way, shape, or form.

The Size of the Sun in the Sky and Other Wondrous Things

I'm sure you've noticed when the sun and the moon are near the horizon, they look much larger than when they are high in the sky. In reality, of course, they are not any larger at all. It's only an optical illusion due to the fact that when they are near the horizon, our brain compares the size of these heavenly bodies to the size of trees and mountains, etc. You know what? It was Ali who showed us it was only an illusion. Ali also said, contrary to guys like Plato, Euclid, and Ptolemy, light doesn't emanate from the eyes; it comes from luminous sources like the sun before being diffused by the objects it hits. In addition to light, Ali was interested in the way the eye perceived light and recognized colors and shapes. He proved—yes, *proved*, it wasn't just some half-baked idea based on a random thought; it was the result of tests and observations—that human eyes capture two images to form a single combined image afterward. How's that for postprocessing? Ali demonstrated—even if this seems completely obvious today, but it wasn't at the time—you can only recognize an object that you're already familiar with. He showed that an image persists in the eye even when you close your eyes, providing proof that memory is involved in object recognition, proving it's more than a question of simple perception. Ali envisioned the *dark room camera*, aka the camera obscura, five hundred years before Leonardo de Vinci had any inkling of the pinhole camera. He thought that the speed of light was not infinite and that light was slower in dense mediums.

Regarding other subjects, he stated the fact that a moving object doesn't stop if a force doesn't stop it, putting him six hundred years ahead of Galileo and Newton regarding inertia. Ali also talked about the way masses attract, leaving us to assume he was aware of the concept of gravitational acceleration.

Bottom line: Ali was strong. Very strong. Buma yé![14]

---

[14] During the historic fight between Mohammed Ali and George Foreman in 1974 in Kinshasa, Zaire, Ali's supporters chanted: *Ali buma yé*, which means "Ali, kill him!" in the Lingala language.

One could ask just why someone like Ali isn't in all the school text-books; there are several reasons, some more pertinent than others. First of all, scientists from the Orient were not exactly celebrities in the West, quite simply because they ended up being translated into Latin, and then they were studied by Western scholars. The work from the Western scholars is what we usually get to see. You'd have to do a lot of historical research—which is scientific research, remember that—to update the genealogy of these efforts. The second reason these scientists are not always recognized is that they usually date back to an epoch when reasoning was often based on ancient misconceptions. Even though the results were sometimes correct, they were overlooked in favor of more recent and more truly scientific scholars. That's the explanation. Whether it satisfies you or not, you'll have to decide. All in all, the *Book of Optics* by Ibn al-Haytham is considered the basis for optical theories developed since that time, right up to the present.

Lens

A lens is usually a transparent component made from a single material. Its purpose is to redirect the light that passes through it. Some say lenses were invented by Italian artisans around 1280 for making telescopes. However, I'd like you to note that the very Honorable Sir Austen Henry Layard, a British archaeologist in the nineteenth century, discovered what seems to be lens-shaped glass in Nineveh, former capital of the Assyrian Empire and part of Mosul in present-day Iraq. The glass is at least three thousand years old.

In 1609, Galileo started polishing some lenses of his own for his research. His full Italian name was Galileo Galilei—his parents were *so* creative!

Now, where were we . . . Galileo was polishing lenses in 1609. He was definitely the first person to use a telescope to look at the sky and make scientific observations. Until then, telescopes were primarily used on ships for looking at things that were far away—not too far away—and sometimes, I'm sure, for watching the hottie next door through the window without being seen. Ahem.

Advances in optics allowed Galileo to develop a geometric

theory on the behavior of light. At the same time, he created a school of thought about light, the "corpuscular" theory. This school of thought would eventually be in opposition to another significant school of thought that arrived on the scene in the seventeenth century, the "wave" theory. To a certain degree, the debate between these two schools of thought has yet to be completely settled.

6. Is Light Made of Particles or Waves?

René Descartes was a French mathematician, physicist, and philosopher born in 1596; his great claim to fame is Cartesianism, derived from his *Discourse on Method*, a true method "for Rightly Conducting One's Reason and of Seeking Truth in the Sciences." His philosophy was based mainly on the idea that humans are able to achieve understanding by reason alone. However, Descartes didn't subscribe to this idea à la Plato or Aristotle, for whom observing the world wasn't necessary for acquiring knowledge. According to Descartes, reason came from intelligence, reflection, and deduction. It also involved memory and the senses and, therefore, observation. That makes Descartes more like Galileo, often considered one of the first modern scientists because of his methodology. In a nutshell, Descartes was known for his rationalism and for his famous *Cogito ergo sum*, "I think, therefore I am." Truth be told, that saying had been around for eons. Descartes aptly completed it, by basing proof of his thinking on primordial doubt: *I doubt*, therefore I think, therefore I am. Descartes is also well known in the world of physical sciences for his work in optics.

Descartes didn't worry too much about the *nature* of light—for instance, its property of instantaneous displacement (i.e., infinite speed). Instead, he studied mathematical laws describing the *behavior* of light, particularly from a geometric standpoint. This is what he wrote in the introduction of his treatise on optics, *La Dioptrique*:

> With another opportunity here to discuss light, [I] only explain how rays enter the eye, and how they can be deflected by the

various bodies they encounter. *It's not necessary for me to try to tell its true nature.* I believe it will suffice for me to use two or three analogies that help in understanding. That seems to me to be the most convenient way to explain the properties demonstrated by experiments. From there, [I] deduce any other properties that aren't so easily observed.[15]

In this treatise, he (re)discovered the Dutchman Willebrord Snell's law of refraction. I say *(re)discovered* because Descartes independently discovered the law, which Snell had apparently also discovered but had never published. Today, we call it the Snell-Descartes law. Not to beat a dead horse, but the Snell-Descartes law should be the Snell law; and honestly, it should be the Abu Saʿd al-ʿAlaʾ ibn Sahl law, or Ibn Sahl law for short, or how about Ali's law? Ibn Sahl discovered it—*and* wrote about it—in his treatise on lenses and burning mirrors in, yes, 984.

So, Descartes published this gigantic treatise on optics. Today, it is still considered the cornerstone of geometric optics. You have to applaud the guy for his willingness to say that he had no idea exactly what light was. Thinking of light as a continuous flux of tiny particles was a good enough approximation for his theories; he honored the scientific method by making it clear that he was describing a theoretical model and if the conclusions can be verified, that's all you can ask for.

Newton wouldn't be quite so careful about it when *he* discussed light.

Naturally, an entire section of this book focuses on Isaac Newton (see page 179). It's completely legitimate to say that one single section isn't nearly enough! No argument here on that. Newton was one of the greatest geniuses in the history of science as well as in the history of people—the species, not the magazine; I'm sure by the end of the book, this running joke will get a smile out of you. Did you know that one, after another, after another . . . many of Isaac Newton's hypotheses were eventually proven *invalid*. More often though, they were amended, usually by Einstein—not too shabby. This doesn't diminish the fact that most of his theories remain valid within their framework. For his time,

15   *Œuvres de Descartes* (Paris: F. G. Levraul, 1824), vol. 5, p. 6; emphasis added.

Newton was about two hundred years ahead of his peers, with all due respect to Robert Hooke.[16] I have to tell you, despite the legend, Newton was never—sorry to burst your bubble—hit on the head by an apple. There, I said it.

Physics Framework

For Isaac Newton, all the laws of physics—governing optics, mechanics, astronomy, and even alchemy—and all laws of nature applied within a given framework, the framework of our reality. Newton described it this way: Space is absolutely fixed, and time is absolute and continuous. This means space is like a stage in a theater; measurements there are the same regardless observer's frame of reference—a meter is a meter in all circumstances. This also means that a second goes by the same way for all observers, no matter where they are, no matter how fast they are moving. This framework, though incorrect, appears to be based on good common sense, even today.

Let's get back to the matter at hand. Newton worked on light, among many other things. He followed up on work by Descartes and Christiaan Huygens, except Newton offered theories on the nature of light.

To be completely honest, I should still mention that, on principle, Newton refused to make hypotheses "freely," meaning he didn't just throw ideas around all willy-nilly. He wrote:

> for whatever is not deduced from the phenomena is to be called an hypothesis; and hypotheses, whether metaphysical or physical, whether of occult qualities or mechanical, have no place in experimental philosophy.[17]

---

[16] Physicist, a contemporary of Newton.

[17] Newton's *Principia* (1687). (The book's full title is *Philosophiae Naturalis Principia Mathematica*, or *The Mathematical Principles of Natural Philosophy*.) Original English translation by Andrew Motte (1729).

Though he remained rather reserved about the nature of light, which was only a hypothesis, it seems that to Isaac Newton light was a stream of tiny particles moving in a straight line in a homogeneous medium. Today, one would say he allowed himself to consider light as a particle as long as this didn't cause a problem with his work.

At the time, scientists were mainly interested in reflection and refraction of light: how light rays propagated in a straight line, bounced off mirrors, and passed through lenses. They wanted to know how to calculate angles to determine light's path. Christiaan Huygens, a Dutch mathematician, astronomer, and physicist, studied optics along with the other scientists, but he theorized that light wasn't made of tiny particles traveling in a straight line; he thought light was waves that propagated in what he preferred to call luminiferous ether, or simply ether. What inspired him to make this assumption? That would be diffraction.

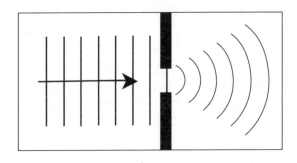

Light diffracted through a slot in a wall

Diffraction was discovered by the Italian Francesco Maria Grimaldi, but it was part of a work that wasn't published until 1665, two years after his death. This discovery was based on a relatively simple experiment. He clearly described it in his work *A Physicomathematical Thesis on Light, Colors, and the Rainbow*. He let sunlight into a dark room through a hole a few millimeters wide, creating a cone of light that projected the sun in the dark room—it's the principle of the dark room camera. He then placed a thin opaque bar into the cone of light and that projected a shadow. Grimaldi found that the edges of this shadow were fuzzy, and that the size of the shadow wasn't compatible with geometrical optics of light as a particle. He also noticed that alternating bands of light and dark had formed. He meticulously measured the bands: width, distance apart, etc. He observed that the color of the bands of light varied, going from violet to white then from white to red as it went from one dark area to the next. He expanded his experiment by seeing what happened when he used differently shaped opaque objects instead of a bar. He went on to add a second hole opening into the dark room.

Grimaldi concluded that the light's behavior could not be explained by existing theories, so a new theory, a new model, was needed. Diffraction was already a well-known phenomenon in fluid behavior, like ripples from a rock in a river. Grimaldi postulated that even if light wasn't a fluid, it acted *like* a fluid. He specifically suggested that the color variations in the light were due to "turbulence" in the stream of light. Grimaldi developed this theory in his treatise, and to say the least, he was on the right track.

Indeed, treating light as a wave was the only way anyone could explain the interference patterns and diffraction. This led Huygens to develop a wave theory for light in his 1690 *Treatise on Light*. Thomas Young experimented on the subject in 1801—you'll hear Young's name in connection with quantum mechanics. In addition, Augustin-Jean Fresnel stated that using wave theory for light was the only way to explain certain phenomena, such as polarization, and the dark areas that appear when two different light sources are combined.

We eagerly waited for James Clerk Maxwell to formulate his equations for electromagnetism. These made it possible to explain the concept of light waves. In 1873, Maxwell offered a definition of light. He demonstrated that geometrical optics, which traditionally used a light particle theory, could be explained using wave theory. You may be thinking, OK then, use the wave model—why was there still such a turbulent debate between the two schools of thought?

Oops, sorry, we're getting ahead of ourselves—you have a few things to wrap your head around before we talk any further about electromagnetism or quantum mechanics. Remember, there are about two hundred years between Huygens's theory and Maxwell's explanation. Well, during those two hundred years, the influence of Isaac Newton had clearly supported the particle theory. I mean, you don't just chuck one of Newton's theories at the drop of an apple—uh, I mean, a hat—not before Einstein anyway!

Newton's corpuscular (particle) theory was heavily criticized when he first published it; the wave theory was widely supported at the time, by Huygens, Gottfried Wilhelm Leibniz,Nicolas de Malebranche, Pierre de Fermat, and Gilles Personne de Roberval—in other words, the vast majority of Newton's fellow seventeenth-century European scientists. However, because Newton had made several discoveries that had revolutionized their concept of the *universe and its behavior*, they understandably decided to give him the benefit of the doubt.

In reality, though, all the fuss wasn't necessary: Newton had never directly claimed that light was a particle, and he never refuted Huygens's work. He had been content to use his theory to simply model light as a particle, for his purposes. He found that his laws for explaining the behavior of light were simple, elegant, and easy to understand. Newton also agreed with Descartes about instantaneous propagation of light, even though the speed of light had already "approximated" in Descartes's day—about which Descartes said, "I'll admit that I am completely ignorant of philosophy, if the sun's light does not arrive at our eyes instantaneously." I'm tempted to say, "so noted."

Newton on Optics

In the area of optics, Newton demonstrated that white light contains all the colors of the rainbow. Before that it was believed that white light encountered particles that carried colors in them, but Newton used prisms to split white light into colors, then recompose it back into white light, showing that white light is truly composed of all the colors of the rainbow. In a way, Newton is the first member of *Pink Floyd*—hmmm, and isn't it interesting to note that pink isn't a real color (see page 61). ✌

Well, everything went along pretty well in the best of all possible worlds; everyone was leaning more toward a wave theory for light, with the possible exception of Newton—the problem being that no one dared contradict Newton for about a hundred years. Now, rewind to what we mentioned earlier: Young, Fresnel, Maxwell. They said light was a wave. So that put everyone in agreement . . . except for Antoine and Edmond Becquerel, Heinrich Hertz, Planck, and finally Einstein. Basically, their journey into the light wasn't over. Far from it!

7. The Discovery of the Photoelectric Effect

In 1839, Antoine Becquerel and his son Edmond Becquerel (that is, Antoine César Becquerel, not to be confused with Antoine's grandson Antoine Henri Becquerel, who discovered radioactivity and shared the Nobel Prize with the Curies, or his *great-grandson* Jean Becquerel, who decided to study the theory of relativity—crazy-smart family, right? Science must be as much fun as a Becquerel full of monkeys!)—well, Antoine and Edmond performed an experiment that brought to light an astonishing phenomenon; when two electrodes were submerged in a liquid, something shocking happened when the lights went on. Allow me to explain: Imagine you have an electrical circuit, like you use to turn on a buzzer, but you don't connect it to any power source. For one section of your circuit, instead of using wire, you stick two electrodes into a conductive liquid. Next, you shine a bright light on the electrodes. Suddenly a current flows and your buzzer buzzes! You take away the bright light and it's over; nothing else happens. Whoa! You've just discovered the photovoltaic effect. That's just what Antoine and Edmond Becquerel did. Happy birthday, Solar Energy!

That was the start of it, but it took a few years—*forty-eight* to be exact—for the experiment to be understood and interpreted by Heinrich Rudolf Hertz. For nearly his entire life, Hertz was fascinated with electromagnetic waves. This German physicist was particularly famous for discovering *Hertzian* waves, aka radio waves—you know, those waves that don't work in your television anymore. In 1887, he discovered the photoelectric effect: When a metal plate was illuminated, it gave off electrons; it gave off more or less electrons depending on the intensity of the light. I simplified the experiment for ease of understanding, but here are some details: Hertz produced an electrical arc that created ultraviolet rays, thereby creating *nonvisible* light; he shone this nonvisible light on a plate of already negatively charged zinc and measured the electric discharge with a measuring device connected to the plate—see why I simplified this? Phew! Hertz's assistant Wilhelm Hallwachs went on to demonstrate that the phenomenon was produced with other metals as well. The experiment was superb—superb enough for a Nobel Prize—but it caused a significant problem directly associated with this effect. The effect occurred only above a certain intensity of light and not at all below that level. If light were actually a wave phenomenon, being exposed to a light with half the intensity would still have produced an effect. Less of an effect, of course, but some kind of effect, right? But there was nothing. It turns out, the photoelectric effect has a threshold (a trigger point); below that nothing happens. Before we talk about Albert Einstein's answer to this problem, let's take a little road trip— we'll stop at Kelvin, pass through two clouds, make another stop at black bodies, and wind up in Max Planck's neighborhood.

8. First Detour: Heat

William Thomson, first Baron Kelvin—hence the moniker "Lord Kelvin"—was an Irish physicist born in 1824. Most of his work and life focused on thermodynamics, the study of heat exchange. He was a fervent defender of atomism, and he spent part of his time studying the relationship between temperature and heat. While that relationship was already well understood by scientists working with gases, it was not as well understood for the other states of matter. As a matter of fact, by 1702, Guillaume Amontons had shown the connection between the temperature and the pressure of a gas: If gas pressure goes up, the temperature goes up; if the pressure goes down, the temperature goes down; in the same way, if the temperature goes up or down, the pressure goes up or down. Kelvin, however, followed on with the work of the French physicist and engineer Nicolas Léonard Sadi Carnot—more on him later—in trying to get away from gases to start talking about temperature and its direct relationship to heat in matter.

We often confuse the two terms, because they are connected; however, some simple experiments demonstrate the difference between heat and temperature. I'm sure you've played with sparklers on the Fourth of July, so you know the little sparks don't burn you when they hit your skin, even though they have a temperature of approximately 1,800°F (1,000°C). On the other hand, if you take a mouthful of hot chocolate, you certainly feel it go down, even though its temperature is probably less than 212°F (100°C). The difference between these two

Heat and Temperature

We usually take the two terms *heat* and *temperature* to mean pretty much the same thing, just like we don't usually think about the difference between weight and mass. Temperature is the measurement of the agitation of the atoms, whether they are free to move about in a fluid—liquid or gas—or held within a structure, such as in a solid. If you measure temperature, you are measuring how fast the atoms are moving in a liquid or how fast of the atoms are vibrating in the solid. Heat, on the other hand, indicates the internal energy available in the matter, energy it can transmit, exchange, or get back in the form of a thermal heat exchange.

experiences lies in the fact that since the sparkler is very small, it contains only a small amount of heat, only a small amount of energy that it can transmit to you, while there's more hot chocolate, so it can transmit a larger amount of heat and burn you. Along the same vein, you may also have noticed, when you heat ice water—with a flame, we fear nothing—the temperature of the ice water doesn't change—nope. The heat received by the ice allows it to change phase, so it melts. The temperature of the water won't increase until all the ice has melted; all the incoming energy is used for the phase change until the ice is gone. Heat is one thing; temperature is another. Time to move on.

Like Carnot before him, Kelvin was interested in the link between heat and temperature; he was sure that Carnot's work would lead him to a way to talk about temperature without being trapped into using laws of physics that concerned only gases—you might say Kelvin had a gas problem. Well, in 1848, he proposed an absolute temperature scale that associated changes in temperature with changes in the heat in bodies (stuff). The scale is absolute in the sense that it doesn't depend on the body being studied, and it isn't based on a reference body—for example, the Celsius scale is based on the freezing point of water at 0°C and the boiling point of water at 100°C. Kelvin's scale assumed a zero value, called "absolute zero," the temperature at which a body contains absolutely no heat, no energy. I won't keep you in suspense. I'll tell you right now, this zero is not attainable—if it were, the incorrectly named Heisenberg "uncertainty principle"[18] would be false, and that would be *bad*,[19] to say the least. Today, the absolute temperature scale is known as the Kelvin scale—and its unit is a Kelvin (K).

When not working on temperature, Kelvin kept busy with quite a few other things; he built an analog mechanical calculator and a machine that could predict the tides. He also calculated the age of the earth; he screwed that one up. We should applaud the effort, though, because he attempted to calculate it with a decent amount of precision. On April 27, 1900, Kelvin gave a presentation to the Royal Institute of Great Britain. In his presentation, he said that people—the species, not the magazine—had already seen everything there was to see regarding physics; the only thing left to do was to make more and more precise measurements. He also said:

> The beauty and clearness of the dynamical theory, which asserts heat and light to be modes of motion, is at present obscured by two clouds.

---

18 One of the fundamental principles of quantum mechanics, which remains valid to this day.

19 Maybe even as bad as "crossing the streams," if you know what I mean. (If you don't, "Who you gonna call?" . . . Still nothing? Well, I guess you can google it. You're welcome.)

The two clouds in question were the inability to prove the existence of luminiferous ether 𝒷 and the problem of radiation from a black body, also known by its more specific name, the *ultraviolet catastrophe*. The first cloud gave birth to the theory of relativity, and the second cloud yielded quantum mechanics, so you see Kelvin was a guy with some serious intuition . . . to say the least.

9. Second Detour: Black Bodies

Max Planck was a rather unusual case in the history of scientists—which is, by the way, almost exclusively filled with unusual personalities. Let's simply say the following: In the 1890s, Planck rejected the atomist theory. He remained convinced that continuity of matter would eventually be demonstrated—that is, he thought it was absolutely impossible to divide matter into elementary building blocks. Nothing unusual about that, since he gladly joined the atomists a few years later, when the existence of the atom had been explicitly proven. What makes Planck particularly unusual is that he put the nail in the coffin once and for all. In one blow, he rendered atomist theory indisputable, laid the foundations for quantum mechanics, and determined the nature of light. Whoa, baby! By studying black bodies.

Black Bodies

We can't go into the technical details— and believe me, there are plenty. Let's just say the following: A black body is an ideal object—therefore, theoretical—that absorbs all of the radiation it receives; it does not reflect and does not send back any radiation. It's sort of like the opposite of a perfect mirror. Because it doesn't send back visible light, it looks completely black—hence the name. However, above certain temperatures, a black body begins to emit radiation.

So what is the problem with black body radiation? The problem is as follows: When you heat a material, such as metal, it emits radiation. This is why metal becomes red, then white, as it heats. In classical mechanics, when you calculate the total energy emitted by the black body, it goes to infinity. Really!? Infinity in mathematics absolutely scares the pants off of any PhD candidate I know. That's how scary infinity is in physics—no kidding, I mean it. It's interesting that there is actually a way to make a black body experimentally: You put an object in an enclosed cavity, drill a minuscule hole—Wilhelm Wien used an oven—so the least little bit of radiation penetrating this body ricochets around in the cavity and can't leave, and it gets absorbed until a natural thermal equilibrium is reached. You can measure the radiation emitted via the orifice while the cavity is being heated. Very fortunately, when they actually attempted this experiment, they observed that the black body absolutely *did not* radiate an infinite amount of energy. Phew!

Actually, the problem doesn't only apply to black bodies. Classical radiation theory predicts that the radiation emitted is proportional to the absolute temperature of the object being heated—thank you, Kelvin—but also inversely proportional to the wavelength of that radiation. Put simply, classical radiation theory predicted that a campfire would emit enormous amounts of killer γ-radiation—the kind that created the Incredible Hulk. Even though the Hulk is a fictional character, it goes to show you the theory's Incredible Error. Classical theory works pretty well for wavelengths in the infrared to green range. However, as you approach blue, or worse yet, *violet*, all hell breaks loose! That's why the Austrian physicist Paul Ehrenfest named this phenomenon the *ultraviolet catastrophe*.[20]

Max Planck starts out with an idea. He was the first one to say he didn't believe this idea at all. He just thought that it might resolve the problem mathematically and provide greater accuracy in predicting radiation emitted by a material being heated. His revolutionary idea earned him a Nobel Prize—Einstein did some more work on the idea and earned a Nobel Prize, too. Here's the idea: And what if the physical phenomenon was *discontinuous*?

[20]   This sounds fantastically dramatic in French—*Cat-ass-trough' ool-trah-vee-o-let!* Say it with feeling—It's the end of the world! What a catastrophe! = *Kell cat-ass-trough'!*

Planck hypothesized that the energy exchanged between matter and the radiation from a black body is quantified, meaning that there's a "smallest" packet of energy that can be exchanged, and that energy can be exchanged only in whole numbers of these "packets." Think of this packet as an elementary unit of energy. If energy were like change in your pocket, this packet would be the smallest coin: the shiny penny of energy. (Take a moment to think about Abe Lincoln and everything he achieved. Done? OK, movin' on!) Planck called this packet a *quantum* of energy—*quantum* means "how much" in Latin. Even though Max Planck himself wouldn't have bet a kopeck on his hypothesis, he had, in fact, just invented *quantum theory*, and that, my dear friends, changed everything.

10. Einstein in 1905: The First Article

In 1905, Albert Einstein worked at the Swiss Federal Institute of Intellectual Property (i.e., the patent office) in Bern, Switzerland. He wasn't in academia; he wasn't known in the scientific world beyond a few friends he got together with to talk about science and eat sausage. However, in 1905, Einstein published four articles that are a tour de force; each article was so important, any one of them could have earned its own Nobel Prize, and they would go on to influence the rest of the world from that time forward. In fact, 1905 is Albert Einstein's *annus mirabilis*.[21]

In the first article, titled "Über einen die Erzeugung und Verwandlung des Lichtes betreffenden heuristischen Gesichtspunkt,"[22] Einstein explained the photoelectric effect using quantum theory. His goal was to demonstrate the existence of a quantum of energy as well as the existence of a particle of light, the photon. Einstein based his theories mainly on Planck's hypothesis from 1900 and explained that light could be emitted or received only by way of indivisible packets—quanta—that move at the speed of light. Regarding the photoelectric effect, he showed that when a light wave was below a certain *frequency*, a very intense

---

[21] Albert Einstein's miraculous year.

[22] "A Heuristic Point of View Regarding the Production and Transformation of Light," *Annalen der Physik*, vol. 322, no. 6 (June 9, 1905).

light didn't produce any electric current, while a low-intensity light wave above a certain frequency did produce light. Ergo, the photoelectric effect had a *threshold frequency*. Einstein explained this threshold frequency by the idea that since a particle of light (a photon) with the correct frequency hits an electron and knocks it off an atom, then only the frequency (color when you're talking about visible light) matters, not the intensity; a single photon will do.

Even though Einstein's theory explained experimental findings remarkably well, there was still a major issue: The theory was in direct conflict with Maxwell's works, which stated all electromagnetic radiation—including light—was continuous and could be divided ad infinitum. Thus, initially, the scientific community had a hard time accepting his idea. Keep in mind, though, that Planck himself, the one who suggested the theory, had a number of reservations. You can see this in a letter prepared by Planck and other physicists recommending Einstein to the Prussian Royal Academy of Science:

> That he might sometimes have overshot the target in his speculations, as for example in his light quantum hypothesis, should not be counted against him too much. Because without taking a risk from time to time it is impossible, even in the most exact natural science, to introduce real innovation.[23]

Despite the doubts and slow acceptance of the theory, Einstein won the Nobel Prize in 1921 "for his services to Theoretical Physics, and especially for his discovery of the law of the photoelectric effect" based on that first article. Even so, the scientific community didn't rally around Einstein until Arthur Holly Compton's X-ray diffusion experiments in 1923. After that, light was no longer considered to be a wave but a particle, a photon.

However, electromagnetism considered light as a wave phenomenon, and the theory worked very well—much better than particle theories. In fact, it looked like wave and particle theory took turns at being

[23]   Official archive of the Collected Papers of Albert Einstein, available at einsteinpapers.press.princeton.edu.

the most suitable model, depending on the phenomenon studied. This issue would give rise to a whole new field of physics: *quantum physics*.

The Photon

The photon is a particle of light—as much as it can be a particle. It's a particularly peculiar particle because along with its ability to act like a particle or a wave, it is electrically neutral and has no mass and no size—so much for being a particle. It moves at the speed of light, obviously; it is *part* of light. Now this is important: a photon can move at *only* the speed of light. It doesn't slow down; it doesn't speed up; and it never stops.

11. Why Do They Turn Off Airplane Cabin Lights for Night Landings?

As you may already know, when planes land at night, aircraft passenger cabin lights are turned off ten or fifteen minutes before landing. Why do they do that? Ask the people around you and you get all kinds of responses! Here are a few particularly good ones: the person to your left might say it's to save energy in case they have to go around again before landing. This one screams—violently—for correction: First of all, can anybody really believe that the energy needed for landing could possibly be affected by a few little lightbulbs? Moreover, why don't they turn out the lights when landing during the day? Because of solar panels? The person on your right might say: It's so the pilot can see the landing strip more clearly. You know, like in a car at night. You don't turn on a dome light while driving, because light reflects off the windshield and reduces the driver's field of vision—yeah . . . uh, no. First of all, the pilot is in the cockpit, which is darkened for the entire duration of the flight for the exact reason of eliminating the light pollution issue—the passenger cabin is completely isolated from the cockpit by a solid door. On the passenger side of the door, Skrillex and Lady Gaga could put on the most spectacular sound and light show ever, and it wouldn't make any difference at all to the pilot. The person behind you leans forward and chimes in: Maybe it's because they want the passengers to be able to enjoy the panoramic view of the city lights at night? Nope. And last but not least—this is my favorite—someone turns around and says: It makes the plane more visible to the control tower. When I hear this reply, I sense this person's intense desire to give an answer—*any* answer, no matter how ridiculous—just to give some kind of answer. First of all, air traffic controllers in the control tower are looking at screens, not the aircraft. Furthermore, illuminating the interior lights would make the aircraft *more* visible in the dark. This last idea is quite simply nonsense—it's not a big deal, but you should know—no scientists worthy of the name would offer that response. The real reason they turn out the lights in an aircraft before a night landing is . . . drumroll please . . . for safety reasons.

To understand why, you need to know how your eyes see. Without going into a lot of detail about how your brain builds a three-dimensional image of what your eyes capture, we'll just focus on the eyes. I'll restrain myself from going off on a tangent about how Aristotle said light emanated from the eyes—in which case, we'd be able to see at night just as well as during the day, and we can't, so the idea was idiotic. Let's start by accepting the idea that any visual information you get is derived from the light that enters your eyes. Whether it comes directly from a light source, or whether the light is reflected from an object, it enters the eye and hits the retina, which is found at the back of the eye. The retina is composed of two types of light sensors: cones and rods—bonus points for creativity if you imagined ice cream cones and pretzel rods when you read "cones and rods."

Cones are more active when there is a lot of light, typically during the day, than when there's not. They sense specific frequencies of visible light, which we commonly call colors. Most human eyes have three types of cones: cyanolabes, which sense blue; chlorolabes, which sense green; and erytholabes, which sense red. When a cone is hit by the right color, it stimulates the cone and sends a nerve impulse—that is, an electric signal—to the brain. The brain then builds an image using the information—but that's not the main point of what we're talking about here. But since we are talking about it, did you know that Daltonians (people with red-green color blindness) suffer from dysfunctional cones? Only two types of their cones function properly. Anyway.

Rods are more active in low levels of light, like at twilight. The rod's job is to sense variations in light's intensity. Rods make it possible to distinguish shapes and movements, but not colors. That's why we see in black and white when there's not much light.

There are two distinct areas in the retina: the *macula*, for seeing what's right in front of you, and around the macula; the peripheral part of the retina is responsible for peripheral vision. The macula has a dimple, the *fovea*, near its center. The macula and the fovea are located opposite the pupil; they sense the part of an image directly in front of the eyes. The macula and the fovea are almost exclusively composed of cones. The concentration of rods increases as you get farther away from the macula and fovea. In good light, you can easily see details when you are looking straight at an object, because the macula is lined up with the pupil. Maybe you've noticed, in dim light your peripheral vision is better, and it's easy to notice small movements from the corner of your eye, as long as the movement causes variations in brightness.

At the front of the eye, we have the iris and the pupil. The iris is the colorful part of your eyes: brown, green, blue—did you know that blue eyes don't really exist? We'll talk about *that* later (see page 59). The pupil is the thick, black disk in the middle—actually, it isn't really a disk at all; it's an orifice. The pupil gets bigger or smaller, dilates or contracts, automatically—it's a reflex—as a function of brightness. In bright conditions, it prevents too much light from entering the eye and hitting the retina; when there's not much light, the pupil opens wide to allow as much light as possible to enter the eye. But in partial darkness, it can't always do the job. Too little light. Too little usable information for the brain. Going quickly from bright light into partial darkness, or from twilight into darkness, makes your brain feel like it's trying to measure the volume of the swimming pool and the volume of a drop of water with the same measuring instrument: In the first case, the drop of water is insignificant; in the second case, the pool completely overwhelms the instrument. The scale of the measuring instrument has to be changed, going from thousands of cubic meters to a milliliter, and vice versa.

The eye has the same issue; it has to make a huge change of scale. How does it do that? As mentioned earlier, the pupil either dilates or contracts to optimize the quantity of light allowed into the eye. It maximizes the amount that enters in dim light and allows a reasonable quantity to enter when the light is bright, because too much light damages the cones and rods, as well as the retina. Light can blind you. Since there are no nerves attached to the retina—other than the optic nerve, which lets you see—the retina doesn't register any pain. As we heard repeatedly in the days just before the American solar eclipse of 2017, never look at the sun without proper protection for your eyes; the sun's radiation could destroy your retina and you won't even feel it.

When you go into dim lighting and the pupil dilates, the brain needs a little time to adapt and change the scale of its perception; this time needed to adapt is called the *dark adaption period*. It takes ten to fifteen minutes for the retina to become relatively well accustomed to darkness, but nearly an hour for the adaptation to be optimal.

Let's run this scenario: You're in an airplane that's just about to land. The cabin lights remain lit. The passenger cabin is bathed in bright artificial light—artificial, yes, but it allows you to see as clear as day—and the aircraft is only a few hundred feet from the ground. Through the window, you see control tower lights ahead of you as well as other lights indicating you are near an airport.

Suddenly, there's a problem! You don't know what's going on. The airplane is shaking violently! Everybody's screaming! Above all the chaos, you just barely distinguish a voice coming over the intercom telling passengers to assume a brace position for impact and prepare for an emergency landing. A few seconds later, there's a tremendous impact! A jolt of pain runs up your spinal column. Your seatback is on top of you, and your face is mashed against the back of the seat in front of you. The plane is on the ground, and you hear the sound of scraping metal as it continues to career forward. You're probably in rough shape, but at least you're alive! After a few very long seconds—You know why accidents seem like they happen in slow motion? We'll cover that too, later, I promise (see page 232)—the aircraft finally comes to rest. Sigh of relief! You're out of danger, it seems. But now, oh no! The cabin is filling with smoke! The lit markers on the floor help guide you to the nearest exit—it may be the one behind you. The exit doors open. The inflatable slides are deployed. A flight attendant who is perfectly calm, in spite of a nasty cut to the forehead, asks you to remove your shoes before jumping on the slide to exit the plane. You take off your shoes—without even wondering where you're supposed to put them. You approach the door and as you are about to jump out—you absolutely *freeze*. No way! You can't jump out. There's no slide! The world outside the aircraft appears to be a dark void; there is only void and darkness.

What's really going on? Well, your retina hadn't adapted to the darkness. You were trying to move posthaste from a brightly lit cabin into the dark night on a landing strip far from house lights and streetlamps that keep night as bright as day. When you put your head through the emergency exit, you and your retina couldn't see: not the slide, not the ground, not other passengers, not even the wing of the plane. Zip. Zero. Nada. Nothing! Would you have the courage to continue? To take a literal leap of faith, because you had been told the slide will be there to land on? Maybe you would. But what about the passengers behind you? And the passengers in front of you? How can our faithful, fearless flight attendant be expected to explain the phenomenon to two hundred panicking men, women, and children who have just undergone more emotional stress in a few moments than most of us will ever experience in our lives? How well will they accept that the standard emergency procedure is to jump out of the airplane into a dark abyss?

I think you'll agree—it's easier to just turn off the lights. The majority of regulations imposed on commercial aviation are implemented specifically for safety reasons: for the safety of passengers, staff, equipment, and surrounding areas. During night flights, safety regulations require that cabin lights be extinguished at least ten minutes before takeoff or landing. This allows the passengers' eyes time to adapt to darkness; in the event of an incident, they'll be able to see where they are going and will be able to exit the aircraft in an emergency. There you have it. That's why they turn off the cabin lights before landing at night.

Interestingly, it's not a new idea. During the days of piracy on the high seas, many a buccaneer wore an eye patch, even though they had perfectly good vision in both eyes. They could rapidly descend to the cargo hold, the cabins, or the gun deck, and by lifting the eye patch, they could see clearly using the eye that was already accustomed to darkness.

In summary, the eye is an extremely sensitive instrument that can sometimes distinguish colors and details, sometimes brightness and movement, and a little of each at any time. Our brain does more than simply adjust our eyes to brightness. Have you ever worn tinted glasses—say, red-colored glasses? After a few minutes, your brain convinced you that what you were seeing wasn't red. When you took off the glasses, your brain didn't see the "filter" anymore; instead, it saw that everything had suddenly become more "blue." Your brain took another few minutes to readjust its perception to the "new" colors it perceived. BTW, everything isn't always only a matter of perception. Speaking of perception, do you know why the sky is blue and the sun is yellow?

12. Why Is the Sky Blue and the Sun Yellow?

This question, Why is the sky blue and the sun yellow? may seem to come out of nowhere, but before we go any further in discussing complicated scientific theories, which often seem to contradict our most basic observations, I think it's useful to demonstrate that our observations are occasionally subject to illusions. Sometimes our observations, while not completely incorrect, are befuddled by complex phenomenon that we just don't understand. Why is the sky blue? Because it's blue, of course! Why is the sun yellow? Because it's yellow, that's all there is to it! But wait, why isn't the sky blue at night? Well . . . maybe you're thinking, "It's because there's not enough light at night to illuminate the sky to show it's blue; so at night, the sky looks black." If that's the case, why can we see stars at night? It would seem to me that the sky has to be transparent for us to see the stars, right? And if it's transparent, why isn't it transparent during the day? And by the way, we say "sky," but it's actually only air. Think about it. When you look at a faraway building, you don't see anything blue between the building and yourself—right?—even though there's plenty of air between you and the building.

Every moment, day and night, the sun emits light every which way into space; some of the light comes toward the earth and arrives here in about ten minutes. The light is white because the sun emits all visible wavelengths—and then some . . . but that's not what we're talking about here. When the sun's rays reach our atmosphere, some of the rays are reflected and sent back into space, like when light strikes water or a mirror. That's why the earth is visible to the astronauts on the international space station: Sunlight reflects off the earth, the way sunlight reflects off the moon, making it visible to us. The rest of the sun's rays enter the atmosphere, then the photons that make up light start crashing into atoms and molecules in the atmosphere. At that moment, the light starts to diffuse, like the light from your headlights in fog. Due to the kinds of gases and particles in our atmosphere, the light with the shortest wavelengths—and therefore the highest frequencies—is what gets dispersed (scattered) the most by this diffusion. In the visible light spectrum, those higher frequencies look blue.

If we want to break down what's going on, the white light from the sun is broken up by our atmosphere and the color blue is scattered, while the other colors continue along on their path with little or no

Rayleigh Diffusion and Blue Eyes

The way sunlight is diffused by the atmosphere as an optical phenomenon was brought to light and finally understood in 1871 by John William Strutt Rayleigh—only one person; I didn't forget a comma or anything—another alumnus of Trinity College, by the way. So what do you think? Do you think the phenomenon that makes the sky look blue could also be responsible for the blue feathers on some birds, the morpho butterfly's blue wings . . . and blue eyes? It's true—blue eyes are not actually blue. Here's what I mean: There is absolutely no blue pigment in the irises of blue-eyed people. When light strikes a "blue" iris, the light rebounds off the iris's structure in a way that only the blue component of the light is reflected back. Isn't that an eye opener!

deviation. The blue light rays striking the atmosphere go off in every direction, and some of it manages to reach our eyes, so wherever we look in the sky, it appears blue. From a strictly objective point of view, you are right about the sky being blue. When I say objective, I mean if we used light sensors in place of our eyes, the light sensors would tell us, yes, the light received is blue. Our eyes are not mistaken. The optical illusion is that the sky looks blue because the light reaching our eyes is indeed blue, even though there's nothing blue between us and the sun.

OK, so, the sky "looks" blue, but why is the sun yellow? Before responding, let's get something straight, right off the bat: The light from the sun is white—I've already mentioned that a couple times. Got it? Now, the sun *should* appear white to us because that's the "color" it emits. The astronauts have assured us that when viewed from space, the sun is white, not yellow. So, why is that?

The answer is quite simple really . . . the sun is yellow because the sky is blue. Voila. 'Nuff said. OK, on to the next subject. No? You want more explanation than that? Well, great! The white light from the sun enters the atmosphere and, as you know, a lot of the blue part of the light is dispersed in the sky. Therefore, the rest of the light, which hasn't been dispersed, continues in pretty much a straight line toward the ground. When your eyes intercept that light, you see the sun that emitted the light, except that you get to see only the non-blue parts of the light—remember, the blue part was intercepted by the sky. It just so happens the visible light spectrum, minus the blue frequencies, makes yellow light. Thus the sun looks yellow to you. There, now you know why the sky is blue and the sun is yellow.

What about sunrise and sunset? The sun looks orange or even red and the sky turns purple and orange. Well, it has to do with how much atmosphere the sunlight has to pass through at that time of the day. When the sun is high in the sky, the light from the sun comes through the atmosphere at pretty much a right angle, which is the shortest distance from the top of the atmosphere to the ground before it reaches our eyes. At sunrise and sunset, however, the angle of incidence (the angle at which the light hits the atmosphere) causes the sun's light to travel through a much thicker slice of atmosphere. This means even more light is scattered during its trip to our eyes—that causes a greater range of colors to show up in the sky. Ergo, the sun is left with the remaining visible wavelengths, the colors with the *longest* wavelengths: orange and red. You may have noticed that I never mentioned pink as one of the colors; not in the sky and not from the sun. There's a problem with the color pink: the color pink doesn't exist. It is another illusion. Pink is actually white light that has lost some of its green component. Pink is non-greenish white (rather than reddish white). Therefore, a pink flower has developed a magnificent way of getting noticed by pollinator insects. It presents itself in a color that is as different as possible from the green grass on the ground: non-green. So now you know, pink isn't a color in the *visible light spectrum*, another name for the "colors of the rainbow."

13. What Is a Rainbow?

When it rains, sunlight passes through drops of water. The rays going through the droplets are refracted by the water—remember refraction (see page 36)? White light from the sun contains all the colors in the visible light spectrum—plus some invisible "colors," but we're not talking about the invisible ones here; stay focused. When white light enters a droplet, each color of light in the white light is diffracted at a slightly different angle, because each color has a slightly different wavelength. So you see, the drop of water acts like a prism—just like on the cover for Pink Floyd's *Dark Side of the Moon* album—only wetter.

Then, the now-separated, monochrome (one color, one wavelength) rays of colored light continue their merry way across the droplet. Depending on the angle of incidence when they reach the other side, they are either refracted again as they leave the drop—*or* they are reflected, like in a mirror, and sent back across the droplet. Due to the geometry of the reflection and refraction, the colored light exits from the same side the white sunlight entered, but at a different angle. Because water is what it is, with its inherent physical characteristics, the colorful light exits the droplet, and the colors spread out from each other, forming a 40° to 42° angle with the sunlight. At that point, according to Isaac Newton, you have red, orange, yellow, green, blue, indigo, and violet. FYI, rather than being separate distinct colors, the colors vary smoothly from one into the next; yielding a multitude of nuances of color—the entire visible light spectrum.

Every moment that it's raining, brilliant monochrome rays of light leave the droplets in a given direction, depending on the position of the raindrop and the angle of the sunlight hitting it. If these colorful rays of light manage to meet up with your eyes, you'll see colorful light coming from the rain. When a raindrop is high in the sky, you see more of the colors that come from lower in the raindrop—closer to red. When a raindrop is lower, you perceive colors that come from higher in the drop—closer to violet.

That covers the vertical aspect of the phenomenon. Next, because light is refracted in all directions by the raindrop, it creates a cone of colorful light. Because of the cone shape, the colored light you see is arriving from the same angle whether from the side or in front of you— actually, any direction, as long as it's in sight and between you and the falling rain. In other words, you are seeing perfectly round, concentric rings of colored light, and the colors go from a red outer ring to a violet inner ring. The light coming from the drops can't pass through the ground, so you get to see only part of the rings in the sky: You are literally observing colored light that is bent or bowed by the rain.

Depending on the angles, it's possible for the light to be reflected not once but several times within the drop. The reflected rays cross and are inverted before leaving the raindrop. This creates double and even triple rainbows. You'll notice that the second rainbow's colors are reversed. The repeating rainbows become fainter with each repetition due to the multiple reflections of the light—reflections are never perfect, and some of the light escapes with each reflection.

So a rainbow is as real as the blue sky and the yellow sun, but it doesn't actually exist. This is an example of an objective illusion, which is particularly striking: You see it, and any appropriate measuring device can sense it, but it's not really there. Vision isn't the only sense that can be fooled. Before we go on to talk about sense any further, I'd like to go over some details you might find surprising: Do you know how many senses you have?

14. How Many Senses Do Humans Have?

Those who could not stop their inner voice from mechanically responding "five!" will now go write on the blackboard two hundred times: "I will not advance Aristotle's asinine assertions." I know, I know, we all learned in elementary school that we have five senses: not four, not six—five! I can even name the five senses for you: sight, hearing, touch, taste, and smell. Where does this exhaustive list of our senses come from? From Aristotle. And how did he go about creating this list? This is where it gets interesting—I think you'll agree it shows just how far off the wall Aristotle, the great philosopher, could get.

From Aristotle, we hear, "We can have no more that the five known senses." Aristotle's proof that we have only five senses? Allow me to paraphrase: we have only five senses because we have no others. Voila! Hey, I hear you, you're saying, "Bruce, when you 'paraphrase' Aristotle, you're actually being sarcastic." So here are Aristotle's own words about touch—translated from Greek, of course—from *On the Soul*. You be the judge:

> If all objects to which the sense of touch applies are actually perceptible to us, any qualities of the touchable object, being touchable, become sensible to us by touch; necessarily, if some sensation of touch is missing, some way of sensing is missing as well. So, *all things that we sense by directly touching them, are sensed by the sense of touch such as we have*; and as for those things which we sense only through some other means without being able to actually touch them, we sense them through the simple elements, meaning by air and by water.[24]

[24] *On the Soul*, bk. 3, pt. 2, chap. 1; emphasis added.

I'll summarize the paragraph: If there are things to touch that we cannot touch, then we lack a sense for touching them. However, we can touch all touchable things. Therefore, nothing is lacking in the sense of touch.

Perfect!

And why five senses? Why not four or six? In *On the Soul*—soon after the quotation just given—Aristotle provides his justification: The five senses are associated with the five elements. More precisely, because there are five elements, there must also be five senses. The five senses—literally the quintessence or "fifth essence"—come to us from Plato and Aristotle; air, earth, water, and fire are the four elements of matter, which form the whole with the fifth. The fifth element, ether, includes everything that is not the other four. Personally, if I'm going to be subjected to this kind of utter nonsense, I prefer Monty Python.

So according to Aristotle, we have five senses, because there are five elements. Here we need a little detour to explain why there are five elements: again, not four, not six—five! Plato tied the existence of five elements directly to the fact that there are only five regular polyhedrons that can fit inside a sphere. A polyhedron, or a solid, is a geometric figure, like a polygon but in three dimensions. A cube is a polyhedron; a parallelepiped—like a pink rubber eraser, or this book (unless you're reading an e-book, but you know what I mean)—is also a polyhedron. A polyhedron is considered a *regular polyhedron* if all of the faces are the same, and if each of the corners are the same. There are five convex polyhedrons called *platonic solids*—even though Plato didn't invent them. If you want to be wildly popular at a party, just bring up these five solids, in ascending order by number of faces: tetrahedron (four faces), hexahedron (aka cube; six faces), octahedron (eight faces), dodecahedron (twelve faces), and icosahedron (twenty faces). If you cut the points off of an icosahedron, you get a soccer ball—soccer is a good antidote for the stress caused by this barbaric nomenclature.

Here's where it all comes together: humans have five senses, because a decahedron (ten faces) won't fit properly into a sphere. And so we are always taught in elementary school that we have five senses.

Since we've been hearing it for twenty-five hundred years, it should be true, right? Or at least, if nothing has been eliminated or discovered in all that time, the five senses are supported by more recent scientific discoveries? Uh . . . nope.

To determine how many senses we have, we need to start by agreeing on just what a sense is—if we don't establish that, people might say sense of humor is a sense, which is nice, but that's false.

So, what is a sense?

Believe it or not, there are still disagreements about the definition of a sense. There is, however, a consensus, and that's where we're going to start. A sense is three things, including, first of all, sensing cells, or receptors, a collection of thingies capable of reacting to an external stimulus—there are disagreements about this point in particular. Next, when the stimulus is sensed, the receptors must generate a nerve impulse to send to the brain. This impulse is called *sensation*. Last, the brain interprets the nerve impulse to provide *perception*. There you have it: the general overview of what the scientists agree on.

Let's consider vision: rods and cones are sensory receptors that sense the color and brightness of light that penetrates the eye; the information received is transformed into nerve impulses—electrical impulses—via the optical nerve. That's sensation. Then the brain collects the information generated by both eyes to reconstruct a three-dimensional image. Ta-da! Vision! The same thing goes for hearing, touch, taste, and smell—because these are undeniably senses. Let's go a little further and look at the human body to see if we can find any other senses. We'll talk about the body in more detail later—consider this a foretaste. Please note: Fore*taste* is not a sense.

Behind the ears—going toward the inside of the skull—along with mechanisms that allow us to hear, we find a small structure called the *vestibular system*, or the inner ear. This system is primarily three small tubes each forming a loop, each oriented perpendicularly to the others, to form what they call in mathematics an orthogonal coordinate system; there is a liquid in the tubes called *endolymph*, or otic fluid, and on the inner surfaces of these tubes you find ciliated cells—cells with cilia, of course; BTW, cilia are hairlike appendages. This is how it works: The endolymph can move freely in the tubes in your head. If you move, or even just turn your head, the acceleration, inertia, and all that—sort of like a bubble level—makes the fluid move. When the fluid moves, it causes changes in pressure on the cilia, the receptors. The information received by the ciliated cells, the sensation, is transmitted to the brain. Since the tubes are oriented in three dimensions, and on both sides of the head, the brain uses the information to precisely interpret your head's movement and position with respect to your surroundings. In a nutshell, this interpretation is our *sense of balance*. This sense has been cleverly named *equilibrioception*. I'm sure you've noticed that while lying on your side, you can see that your environment is turned 90°, but you still know full well which way is up and which way is down. On the other hand, have you ever watched video from a camera that was turned 90°? It's impossible to watch the image normally without tilting your head, even though you know where up and down are. In the first case, your brain tells you where the floor is. In the second case, only your sense of vision is working—and if you watch your sideways video with your head vertical, your brain tells you, "Wrong, the ground is actually down there." It's the same as when you are in a boat cabin, and your vision tells you that you are in a furnished room, but your equilibrioception says you are moving with the boat as it rises and falls on the waves. That means your brain is receiving conflicting information from two sources and your brain does *not* like contradictions. If your brain is forced to handle conflicting information, maybe it will, or maybe your brain will decide that your eyes need to see the last thing you ate, to put it delicately. That's why seasick passengers are encouraged to go on deck and look at the horizon: This way they can see that the boat moves, which creates agreement between the two senses. Equilibrioception, the sense of balance, is therefore a sixth human sense—nanny-nanny-fou-fou, Aristotle. And there's more to come!

There are two additional senses which some people—the species, not the magazine—consider to be part of the sense of touch. Every time I talk about either of these senses, someone tells me that I'm simply talking about touch, and it's not actually a separate sense. Having heard that, I hope you are sufficiently curious that you'll wait a moment or two before telling me that perception of heat and perception of pain are merely specific sensations of touch. Thank you. Now then, there's nothing particularly complicated about the way these two senses work. *Thermoception*, or the perception of heat, depends on two forms of sensory receptors that are present in most mammals: heat sensors for heat, and "no heat" sensors for cold. Let's not talk about "internal" heat sensors—meaning those that sense internal body temperature—so we can concentrate on external heat sensors, the TRP[25] family of proteins, like vanilloid and melastatin. Interestingly, some animals are equipped with infrared detectors, so they can literally "see" heat, but that's not the case for humans. In humans, TRP proteins are activated by changes in temperature. Many of these sensors are located on the skin; this, of course, explains why many consider sensing heat to be part of the sense of touch. However, one of these proteins exists *other* than on the skin. It allows us to definitively differentiate thermoception from touch. That protein is TRPM8.

Why Is Menthol Cold?

Protein TRPM8, or transient receptor potential melastatin 8, is a cold-sensing receptor on the tongue. It is activated by temperatures between 59° and 82°F (15° and 28°C). This protein, composed of more than 1,100 amino acids, is also a channel for calcium; this means that it is also activated when calcium ions bond with it. That might seem like an unimportant detail until we note that menthol activates calcium ions. Thus contact with menthol activates the cold receptors, and a sensation is sent to the brain; the brain can't tell what activated the sensors, so it's happy to just interpret the nerve impulse and say "cold." That's the reason menthol is cold. And you must agree, in this case, it's certainly not a matter of touching.

Nociceptors, or pain receptors, detect pain. They play a crucial role. Without them, it would be impossible to immediately detect dangerous situations like touching a hot oven. Some people suffer from a congenital

[25]   Transient receptor potential.

insensitivity to pain. It's usually a genetic condition and makes them absolutely indifferent to pain over their entire body, with no loss of the sensation of touch—I *told* you touch and pain were independent. At first glance, life without pain may seem attractive, but you better believe it's not. Their life expectancy is much lower than the life expectancy of other folks, because someone with congenital insensitivity to pain risks being continuously exposed to severe trauma and health issues that go unnoticed. Imagine breaking your arm and not feeling it. You proceed calmly through your daily routine without doing anything about your arm; you just continue using it as normal. Imagine putting your hand on an electric stove burner, and then, after a few minutes, you notice the horrible smell of burning flesh without ever considering it might be your hand. Surviving infancy must be incredibly difficult for a baby who can't tell the difference between sucking its thumb and biting it. *Nociception*, the sense of pain, is an absolute necessity for the survival of the human species. It would be a pathetic underestimation to classify such an important sense as a subcomponent of touch, just because our skin is covered by nociceptors. Think about it; they aren't only on the skin. Have you ever experienced a toothache? A headache? A stomach ache? A muscle cramp?

Maybe you're someone who thinks we should link nociception and thermoception—at least that's what I'm about to do here—because a burning sensation is the simultaneous sensation of heat *and* pain.

Why Do Hot Peppers Burn?

Regarding heat, the great family of TRP proteins includes members that are also pain indicators; such is the case with protein TRPV1, transient receptor potential vanilloid 1. This protein is activated at temperatures above 111°F (44°C) or at a low pH, like in the presence of an acid, which is the reason why an acid "burns." In reality, acid and heat do not do the same things at all, but your brain interprets them the same. The TRPV1 proteins are heat-activated; they are also activated in the presence of capsaicin. Capsaicin is a molecule found in hot peppers, and that, my dear friends, is why hot peppers burn, baby, burn!

If we consider nociception and thermoception as two separate senses—and believe me, we must—we now have *eight* senses. Dear Aristotle, why couldn't you have come up with two senses per element? You could easily have discovered nociception and thermoception; we know that you actually asked yourself about the nature of heat.

Now, let me ask you a bizarre question: Do you know where your feet are without looking? Not in absolute terms; I don't want some clever answer like "at the end of my legs." Do you know where they are in space? More precisely, if you close your eyes, can you point at your feet with your finger? Go ahead, try the experiment, but only if no one's watching you—otherwise, you look like someone who's giving their foot a real talking to. Consider this: You know where to put your feet even when you're walking in the dark—same goes for any part of your body; they don't do it themselves. This is one of those things we never think about. Imagine if you couldn't find something in your car or backpack by feeling around for it. Imagine if you didn't always know the location of your body parts when you weren't looking at them. *Proprioception*, or *kinesthesia*, is the perception of your body position. To convince yourself that it is indeed a sense, let's look at what happens when it's not present.

Deafferentation is a problem that is as terrible as it is rare—there are currently four documented cases in the world. A deafferentated person loses all sensitivity in their limbs—due to absence of information coming from *afferent* nerve pathways. These are the nerve pathways responsible for sensing external stimuli, as opposed to *efferent* nerve pathways that are responsible for controlling movement. A person with deafferentation can move; she has complete use of her entire body but no sensation. In particular, she completely lacks proprioception; she has to use her vision to make up for it. For example, if someone serves you a cup of coffee, your vision locates the cup of coffee in space, then you grasp it with your hand. You didn't have to look at both your hand *and* the cup the entire time. Every instant that your hand is moving toward the cup, your brain is evaluating the position of your hand in space. This evaluation happens so quickly it seems instantaneous, but your brain is

processing the nerve information sent from your hand. A person suffering from deafferentation can't do that. If she loses sight of her hand for
even an instant, she no longer knows where her hand is. If she looks at
her hand and loses sight of the cup, she's unable to tell if her hand is
moving in the right direction to grasp the cup. Any movement she
makes has to be guided by sight: walking, sitting, getting up, brushing
her teeth, putting food in her mouth, brushing her hair, getting dressed,
etc. In the dark, a person with deafferentation has no way of knowing
exactly what she's doing with her limbs.

Proprioception involves all the afferent nerve pathways combined,
for all transmission of sensation. As for the sensors, all sensors in the
body participate in the sensation: pressure (touch), heat, pain, balance,
etc. Surprisingly enough, one of the most critical senses for our normal,
everyday functionality is also one of the least known. Proprioception
comes up as the *ninth* sense in our list.

Are there any others? According to some scientists, yes. There's the
sense of hunger, chronoception (sense of passage of time), electrocep-
tion and magnetoreception (sensitivity to electric and magnetic fields),
echolocation, etc. Some scientists estimate that humans possess no
fewer than twenty-one senses; however, they are nowhere near having a
consensus on the subject. At present, regarding most of the "additional"
senses, scientists are still trying identify the sensors and how they are
activated—basically, they still don't understand how these senses work.
I've listed them here as recognition, the way you recognize someone
gave it his all to capture fourth place in a competition. Until further
notice, we'll keep the total at nine senses. The other senses remain
hypotheses.

All in all, humans possess nine senses: sight, hearing, touch, smell,
taste, balance, thermoception, nociception, and proprioception.
Remember earlier, I mentioned sight isn't the only sense that can play
tricks on us (see page 70)? Now that you know a little more about the
senses, let's talk about touch. Regardless of what you may think, you
have never touched—nor will you ever touch—anything in your entire
life.

15. **You Have Never Touched Anything in Your Life**

As you read these lines, you are resting on something, maybe sitting in a chair, lounging on a sofa, lying on a bed, or stretched out on the grass. You must be touching something. We might assume that if you are sitting in a chair, your body is touching the chair. At a minimum, your clothing is touching your body. But are they *truly* in contact? First, let's define *contact* as having no empty space between two bodies, then let's examine the situation down at the atomic level. Now, we ask, are they really in contact? MC Hammer brings us the shining answer: "U Can't Touch This!"[26]

The answer is no. Contact is quite simply impossible. At the atomic level, you'll remember, atoms have electrons around them. If we consider atoms to be tiny spheres, the electrons are the outer layers. Also

Things Can't Touch on an Atomic Level? Really?

In reality, things are a tad more complex than that, as we will see when we discuss quantum mechanics. The fact is, at an atomic scale, electrons can't be considered as "tiny spheres." Therefore, you can't define contact between two electrons the same way you would define contact at our macroscopic scale: When my two hands are in contact, I cannot pass a sheet of paper between them. At the atomic scale, that has no meaning. In addition, because electrons are always moving around the atomic nuclei, when two atoms are "side by side," the displacements of the electrons, at a given moment, can create an attraction between atoms—the positions of electrons around one atom can make a variation in the positive charge on the side where the other atom is located . . . that kind of thing. The definition of contact at the quantum scale can be defined as the distance at which the forces of attraction and repulsion between two atoms are balanced. This, of course, does not mean the same thing as contact at our scale. Our kind of contact is forbidden at the quantum scale.

[26] *Please Hammer, Don't Hurt 'Em*, 1990.

remember, electrons are negatively charged. As your skin gets closer to the chair, the atoms and molecules in the outermost layers of your skin get closer to the atoms and molecules in the outermost layer of the chair—with their electrons leading the way. Eventually, the electrons get so close to each other that their electric charges begin to interact. Since electrons obey Coulomb's law, they repel each other. The bottom line is: Matter cannot permit contact—other than in extremely violent cases, such as atomic fission. Even then, strictly speaking, they don't completely meet the conditions for contact.

Perhaps you're thinking, "But it is possible to force the matter of one body through the matter of another body. For instance, when I push a pencil against a piece of paper, the pencil goes through the paper." Nope—even then—no contact. Here's what happens: At the very moment when the pencil's electrons repel the paper's electrons, you force the atoms of the pencil forward; by doing that, you force the atoms in the paper to separate.

This is due to an interaction I've already mentioned in this book. It is also a phenomenon we encounter on a regular basis. Believe me, when I say "regular basis," I mean during every fraction of every instant of every second of every moment of every day of every life. This never-ending interaction goes by the name of *electromagnetism*.

ELECTROMAGNETISM

Of magnets and lightning.

16. Magnetism

Since ancient times, mostly likely in China first, then in Greece, humans noticed a mysterious kind of stone that had the magical power to attract iron. Amazingly, this deep black, sometimes extremely shiny stone could also transmit its magic power to iron. Once "enchanted" by the mystical stone, iron could also attract iron. The stone was called "loving stone" in French because it was so affectionate and *lodestone* (leading stone) in Middle English, because it always pointed to iron. It eventually came to be known as *magnetite* from the Greek, μαγνητ, meaning "stone of Magnesia." The name is derived from Magnesia, the name of an ancient city, in what is now present-day Thessaly. Magnesia was such a great place, even the rocks were attractive!

Around the year 1000 CE, the Chinese first noticed that a "magnetized" (in quotes because they had no idea what magnetism was or how it worked exactly) iron needle could be useful for navigation; the needle always pointed in the same direction. About two hundred years after that, the compass was discovered—independently—in Europe. Then a few years after it was discovered in Europe, a French scholar, Pierre de Maricourt, started studying magnets, their properties, and the way they work.

Maricourt wrote a formidable treatise on magnets with the sober title of "Epistola Petri Peregrini de Maricourt ad Sygerum de Foucaucourt, Militem, de Magnete,"[27] more commonly known by its shorter title "Epistola de Magnete"—not to be confused with *De magnete* by William Gilbert and his partner Aaron Dowling; the actual title they used was *De magnete,*

---

[27] "Letter from Pierre 'The Pilgrim' de Maricourt at Sygerus de Foucaucourt, Soldier, Regarding Magnetism," 1269.

magneticisque corporibus, et de Magno Magnete Tellure.[28] In "Epistola de Magnete," Maricourt flawlessly constructed the laws of magnetism⍟ in the first part. But in the second part, he looks at the possibility of using magnets to create perpetual motion—which is impossible. Just so you know, there is a lot of idiotic stuff on the internet about perpetual motion. A dismaying number of those sites claim this stuff was invented by Nikola Tesla. Poor Nikola, now he's *perpetually* turning over in his grave.

Perpetual Motion

Newton showed that it is *theoretically* possible to create perpetual motion, but there is definitely a difference between theory and practice (*practice* is an efficient way to say, "what can be achieved"). In practice, perpetual motion is impossible. Perpetual motion is motion that won't ever stop unless an outside force stops it. Over the course of history, many have tried to create this kind of motion—and they have failed.

Way back when, I'm sure it seemed reasonable to consider magnets as a way to achieve perpetual motion. In that era, when a magnet moved a piece of iron or another magnet, it looked to them like motion without force, or to say it more correctly, it wasn't necessary to apply a force to get it to move. And if you imagine an magnet sliding on a track with a magnet at each end to repel it, you might think that the sliding magnet would be repeatedly repelled by one magnet, then the other. It would look a hell of a lot like perpetual motion. It doesn't work, though, because (1) the sliding magnet rubs against the track, so friction slows it down with each trip until it finds a point of equilibrium along its path, and (2) a magnet doesn't last forever; it loses its magnetism, usually after a few centuries, but sooner with sharp impacts or at elevated temperatures.

Sorry to tell you this, but you know those "magnetic engines the oil companies and automobile manufacturer lobbyists are hiding from you"? They don't exist.

You can't really blame Maricourt for trying to use magnets to achieve perpetual motion. It wasn't until after his era that we

---

[28] *On the Magnet, Magnetic Bodies, and the Great Magnet Earth*, 1600.

learned—thank you, laws of thermodynamics—perpetual motion is just absolutely not possible.

Maricourt presented the fundamental laws of magnetism in a compellingly logical order. He began by giving magnets a north pole and a south pole. He chose those words, because the two ends of a magnetic compass needle always pointed toward the earth's North or South Poles. Using two compasses, he found that if the ends of two compass needles point toward the same pole, they repel each other, and the ends that point at opposite poles attract each other. From that he deduced that the south pole of the compass needle points toward Earth's North Pole and vice versa. Pretty clever. He also told us that it is not possible to separate the poles of a magnet—like by breaking it—without creating two new poles at the point of separation.

Maricourt didn't understand that the earth itself was magnetic. He thought, rather, that magnetism emanated from the celestial dome of the heavens, the sky. So he converted the compass into a ball-shaped magnet with clearly marked poles; he thought that if the ball was free to move in a liquid, it would be able to track the rotation of the earth. He expected its meridians (longitudinal lines) to remain aligned with the celestial meridians they represented. Here is where he thought he'd achieved perpetual motion—in this case, in the form of a clock. But even though he didn't get perpetual motion out of it, he did improve the compass so that it was much more accurate. And a few years later—230 years later to be exact—the compass helped Christopher Columbus completely fail to reach Japan and instead discover the Caribbean Islands—because, yes, he was expecting to stop in Japan before going on to the East Indies . . . and no, he didn't discover America. So Maricourt understood all those things about magnets. That's all well and good, but we still don't know how magnets work.

17. Permanent Magnets

"Permanent Magnets" is a sexy title, but as you already know, magnetism isn't very complicated . . . as long as you don't dig too deep. And we're not going to dig too deep, at this point anyway. Let's simply start with the idea that an electron behaves, in some ways, like a tiny magnet. And, OK, let's say a proton acts like a tiny magnet, too—one that is much, much, much weaker than an electron.

Magnetism is interesting, because as commonplace as it is, it's one of those rare cases in which we can experiment with quantum phenomena on our own scale. Electrons are, among other things, like tiny magnets with a north pole and a south pole. Depending on an atom's configuration—specifically, the number of electrons and how they are distributed on the atom—the effects of the tiny magnets may or may not cancel each other out.

Each electron creates its own aura of "magnetic magic," a magnetic field. A magnetic field is a region in space—very close to the electron, in this case—that can be influenced by this "magnet." The combined effects of each electron's orientation in the atom can sometimes create what we call an atomic magnet, a teeny-tiny magnet. It's the size of an atom and it's more than just electrons this time. This time we find ourselves in the company of an atom that forms a little magnetic field around it. This is the first and smallest scale of magnetism.

But when you have a bigger bunch of magnetic atoms, like maybe iron or magnetite atoms, you have several possible cases: (1) the atoms may be randomly oriented; (2) they may all be perfectly oriented in the same direction; or (3) they may be locally oriented, in patches, with some atoms oriented in the same direction in one area, and some atoms oriented differently in another area.

And on a larger scale, what determines the orientation of the atoms? Say a rock-size bunch of atoms? The atoms arrange themselves in the way that needs the least amount of energy to be stable—because nature is a heinous slacker.

In the first case above, with the magnetic atoms oriented in all different directions, the macroscopic stuff formed by these atoms barely generates any magnetic field; thus the stuff isn't magnetic. In the second case—you are about to appreciate the compellingly logical order of the presentation—when the atoms are magnetically aligned in the same direction, at a larger scale they behave as if you've lined up a bunch of magnets with the north and south poles end to end. In this case, the stuff is magnetic, like magnetite, for instance. In the third and final case, finally—yes, finally—in the third case, the magnetic fields cancel; if the stuff is made of oriented atoms found in patches here and there, it isn't magnetic. It doesn't create a magnetic field.

But back in the mists of the time, this is what seemed so magical: When they took a piece of iron, they could see that not only was it attracted by magnets but it could become magnetic itself. What really happens when you bring magnetite and iron together? If it's strong enough, the magnetite reorients the atoms in the iron. That means magnetite can force all the atoms within its sphere of influence to magnetically align in the same direction; iron falls into the second case. It's a material that can be magnetized.

It's breathtaking! At our macroscopic scale, we can witness the manifestation of a quantum phenomenon—said to be microscopic and beyond our grasp. Ummm, yes, I do realize that I have merely postponed the issue by presenting the electron as a tiny magnet without explaining why—but we're taking baby steps on the way to quantum mechanics 𝔚.

You might ask why the title of this section is "*Permanent* Magnets." Well, it's because there are two kinds of magnets. Some magnets are magnets all the time—gee, well, like a magnet—and some magnets are magnets only when you run an electric current through them: electromagnets.

18. Electricity?! What Does That Have to Do with Anything?

Before going off on another little detour through the history of the subject like we did at the beginning of the book, I'm going to start by explaining a little bit about what electricity is. Why am I allowing myself this luxury? Two reasons: First, I'm doing the writing, so I do what I want; second, because you've already heard some history on electricity —remember amber and all that?

We're going to take the scenic route, as you'll soon see. We'll start with what we already know about electrons and protons (they have an electric charge) to explain electricity. I'm going to define electricity with electricity—my book, my rules—just like I did with magnetism. Do you remember when we mentioned that opposite sign charges attract and same sign charges repel? Electricity is an effect, the result of the motion of particles when they are charged, whether they attract or repel. Some materials, *conductors*, have a structure that allows the electrons of its atoms to move from one atom to another more easily.

Imagine a chain of atoms. Let's call it *conductive*. At the beginning of the chain, add an electron. Now, the first atom has one electron too many![29] Next, the first atom gets rid of one electron—not necessarily the same one—and pawns it off on the next atom in the chain. And so on, and so on. In summary, one electron from each atom moves to the next atom in the chain. And what happens at the end of the line? If the chain is in contact with a material that doesn't conduct, an *insulating* material, then the last atom can't get rid of its extra electron. *But*, the moment that chain gets near a conductive material, the extra electron is immediately thrown off and transferred to the next available atom. We might also think—theoretically—that if we *pull* an electron off the first atom, it would steal an electron from the next one, and so on. What we deduce is that charges can move in either direction in a conductive chain. And we are right!

Great. That's all well and good. We have a stream of electrons moving along, a little like water running through a pipe, but what's the relationship between electricity and being able to light a light bulb? Well, there's more than one way to use electric current to generate light, but *currently*, the most common method uses a material's resistance to electron displacement. That would be the incandescent light bulb, the one with a filament that heats to a very high temperature—the one that *was not* invented by Thomas Edison, in spite of the legend of his tremendous tenacity.

In an incandescent light bulb, its filament is part of an electric circuit, meaning electrons will move through the filament. The filament material, even though it is a conductor, is *a lot less* conductive than your

[29]   Because, yes, an atom must always have a well-defined number of electrons.

typical electric wires. Therefore, electrons have "a traffic jam" trying to get through the filament and they "bang into" their atoms. Actually, of course they don't really hit anything at all, but obviously, there is agitation and vibration; this vibration heats up the material—an agitated structure is hot. The filament gets so hot that it emits light—it becomes incandescent. The special gas around the filament might diffuse some of the light from the filament, but mostly the gas surrounding the filament keeps it from burning up.

Alessandro Volta, an Italian physicist, was keenly interested in electricity and a machine called an electrophorus, which generated static electricity (see page 80) by rubbing amber against fur or silk against glass—you get the same result either way. In 1775, Volta improved the design; the electrophorus was originally described in 1762 by Johan Wilcke, a Swedish physicist. Later some work by an Italian physician, Luigi Galvani, caught Volta's eye. Then in 1781, Galvani discovered a phenomenon he called "animal electricity." He used a frog's leg to connect two metal disks, each made of a different metal. WHAT?! The frog's leg contracted, which proved there was an electric current—well, it worked didn't it?

Volta reproduced the experiment using cloth soaked with very salty water—brine from the ocean—instead of a frog's leg. He found that electric charge was exchanged between the two metals. Volta then performed the experiment with all different kinds of metals. He determined that the electric potential between metals depended only on the kinds of metals he used. And he found out nothing happened when he used the same kind of metal on each side of the brine-soaked cloth.

In 1800, he continued work on his experiment and realized he got excellent results with zinc and silver. He stacked a pile of plates, alternating zinc and silver, and each pair of plates was separated by a brine-soaked piece of cardboard. A stable electric potential appeared at the ends of the pile. This pile is called a *voltaic pile*. It's the ancestor of the electric battery: Volta's battery, or the *voltaic* battery. Today, electric potential, aka voltage, is measured in volts in honor of Alessandro Volta.

A few weeks later, while reproducing Volta's battery-building experiment, two British chemists, William Nicholson and Anthony Carlisle, discovered something pretty astounding: When they ran wires—we now call them electrodes—from the ends of the battery down into water, they saw bubbles forming on the electrodes. Because they were aware of all those earlier advances in chemistry that we just learned about, these two chemists were able determine the composition of the gases formed at the electrodes. They found out that oxygen appeared on the anode, the one attached to the positive end of the current generator, and hydrogen was produced at the other end—the cathode, the negative one. Nicholson and Carlisle realized they had just "broken" water molecules using an electric current—electrolysis of water. And by the way, they also just happened to prove it was possible to convert electrical energy into chemical energy!

19. Static Electricity

Let's talk about static electricity—historically, the first electricity discovered—what you get when charges don't flow through a conductive material. It's what you get with amber. When you rub amber on fur, amber pulls atoms off the fur's atoms. You produce the same phenomenon when you shuffle your feet on the carpet. Your feet pull electrons off the surface of the carpet and your body charges up with electricity. Since the carpet is an electric insulator, you can't give those electrons back to the carpet. The accumulated charge waits—oh, so patiently—until you come into contact with a conductor so it can—ZAP—discharge! That's what happens when you eventually touch a metal doorknob or another person. The electrons are looking for the shortest path to get from your body to the doorknob, so they leave you by jumping across a short distance through the air. It causes a small but extremely annoying electric spark. Lightning is a manifestation—on a much, much, much larger scale—of the same phenomenon. They are both static electricity.

When a particle is electrically charged, it interacts with another electrically charged particle repelling or attracting it, depending on whether the charge is the same sign or not. Charles-Augustin de Coulomb took advantage of this when he built an ingenious device to

Lightning

In a cumulonimbus cloud—a storm cloud—there are large temperature differences between the top and bottom of the cloud. This makes a lot of mass move within the cloud. This motion, driven by temperature differences, is called *convection*. But due to very low temperatures in some places, you find hailstones and sleet—tiny pieces of ice—along with the dust suspended in the air within a cumulonimbus. The motion due to convection causes different things in the cloud to rub against each other and causes a triboelectric effect, aka *triboelectric charging*—awesome Scrabble word—the technical term for rubbing a conductor against an insulator such as amber or fur. So, we have a lot of stuff working up quite a charge in the cloud. And the cloud's composition makes it polarize quickly and completely; 90 percent of the time, the positive charges collect at the top and the negative charges at the bottom—the inverse happens only 10 percent of the time. Now then, a very strong electric field has developed around this cloud—the tension is killing me; we know what's coming. The surrounding air, like any insulating medium, has a *dielectric strength*—that is, a threshold of tolerance for electric fields. Above that threshold, an electric arc is produced, meaning the charge finds a path. When the cloud is sufficiently charged and exceeds the dielectric strength of the air around it, the cloud discharges through the air to the ground.

Here's how it goes. First, the discharge snatches all the electrons from the surrounding air, creating a positively charged channel through the air. Then, a small portion of the charges from the cloud propagate (move along) at a speed of 124 miles/second toward the ground, branching here and there based on the local air characteristics—pressure, wind, temperature—and the energy needed to cross through it. Nature always chooses the path that requires least energy. Below the storm, near the soil, positive charges are accumulating—mainly due to how close the ground is to the storm—and a similar process starts, this time, from the ground toward the sky. When the descending channel, "a leader," and the ascending channel, "a streamer," meet—these channels can't be seen before contact—there is a complete discharge, "the stroke," as they call it, and an electric arc appears—lightning!

The charges move through this channel at speeds up to 62,000 miles/ second. The channel guiding the lightning can be up to 15.5 miles long, but only 1.5 inches across. The discharge creates a plasma—the fourth state of matter, not the blood product. As the plasma dissipates, it produces the flash (that's the lightning) and a sonic boom (that's the thunder) caused by extremely rapid expansion of the air.

measure static electricity by measuring how strongly two electrically charged metal balls attract or repel each other. This machine, called Coulomb's torsion balance, allowed him to figure out how charges interact, thus allowing him to present his famous law, Coulomb's law—the one we talked about earlier to explain that same-sign charges repel and opposite-sign charges attract.

What's key is this interaction happens from a distance, without contact. It means an electric charge has some sort of sphere of influence. Coulomb showed that wherever static or dynamic electricity exists, it influences the space around it. This sphere of influence is called an *electric field*. When the field's charged particles are static, we talk about an electrostatic field. When the field's charges are moving about, we call it an electrodynamic field.

20. Electric Fields

We humans have been aware of electric fields for as long as we've been aware of static electricity. In fact, charged amber attracting tiny bits of stuff seems so magical specifically because it happens at a distance without any need for contact. Ooooo! Ahhhhh! (Actually, it really is pretty cool.)

Through the eighteenth century, many scientists studied magnetism or electricity or both. Way back in 1600, in his famous *De magnete*, William Gilbert suggested using the word *electrick* for talking about electrostatic phenomena. But it was really Hans Christian Ørsted who "first" demonstrated the link between electricity and magnetism. In 1820, he demonstrated this link by means of a very simple experiment on electric fields. It's so famous that almost every high school student has seen it. The experiment is even named the Ørsted experiment. BTW, the experiment had been performed eighteen years earlier in Italy, but the 1802 discovery had been met with general indifference. ·

In April 1820, Ørsted put a compass near a wire—close but not touching said wire. When there was no electric current, the compass indicated north-south as usual. But when electric current was flowing, the compass needle consistently indicated a new direction—always the same new direction. The experiment was reproducible. Anyone could put a compass next to a wire and watch the needle move as the current was turned on and off. If you kept the compass and wire in the same spot, the needle always moved to the same spot based on the current flow. Ørsted refused to see the link between electricity and magnetism; he thought they simply interfered with each other. As legend has it, while performing the experiment, his lab assistant felt sorry for Ørsted, because he thought the moving compass needle meant something wasn't working properly—showing that a discovery can be made only by an open mind.

Some weeks later, in July, Ørsted published his results in Latin—under the title *Experimenta Circa Effectum Conflictus Electrici in Acum Magneticam.*[30] Ørsted himself acknowledged that he was familiar with previous work by Gian Domenico Romagnosi and said that it contributed to the discovery of electromagnetism. Eighteen years before Ørsted, the Italian magistrate-philosopher-economist-physicist Romagnosi had performed the exact same experiment in Italy and discovered the magnetic effect of electricity. He had even published articles about his discovery in Italian in local newspapers—where it went rather unnoticed—and he submitted his discovery to the French Academy of Sciences. Oh, I should mention . . . he lived in Trento. And at the time, Trento was occupied under Napoleon. Romagnosi's findings were probably ignored because he was a magistrate by profession. To those academy physicists, having scientific advances made by a magistrate was even worse than scientific advances made by a geologist. Wouldn't you agree, Mr. Chancourtois?

---

[30] *Experiment on the Effect of Electric Conflict on a Magnetized Needle,* July 1820.

To wrap it all up: To give you *some* idea of what an electric field is, and to give you a *good* idea of what electrical force means, I think it's a *great* idea to quote one of the greatest physicists in history and absolutely the best science educator of the twentieth century—possibly of all time—Richard Feynman!

> Consider a force like gravitation . . . but that is a *billion-billion-billion-billion* times stronger.[31] And with another difference. There are two kinds of "matter," which we can call positive and negative. Like kinds repel and unlike kinds attract. . . . There is such a force: the electrical force. . . . So perfect is the balance, however, that when you stand near someone else you don't feel any force at all. If there were even a little bit of unbalance, you would know it. If you were standing at arm's length from someone and each of you had one percent more electrons than protons, the repelling force would be incredible. How great? Enough to lift the Empire State Building? No! To lift Mount Everest? No! The repulsion would be enough to lift a "weight" equal to that of the entire earth![32]

That's the kind of force we're talking about. And did you notice? We've bridged a gap between electricity and magnetism! From here, we'll need a few more things and a few more people to help us understand that electricity and magnetism are actually two facets of one natural phenomenon: *electromagnetism*.

21. Ampère, Gauss, Faraday, and Others . . . Right up to Maxwell

André-Marie Ampère was a French mathematician, physicist, chemist, and philosopher. He studied Ørsted's 1820 experiment very closely. Using Ørsted's experiment, along with other works that preceded Ørsted's,

---

[31]　He's not exaggerating the scale of magnitude!

[32]　Richard Feynman, *The Feynman Lectures on Physics*, vol. 1, 1964, available at feynmanlectures.caltech.edu/II_01.html.

Ampère built a comprehensive theory of dynamic electricity, aka electrodynamics, in addition to his in-depth research into magnetism.

Ampère focused on electric current, its direction and sign. It was Ampère, by the way, who showed the direction of the current was based on completely arbitrary conventions; when a current exists, negative charges can move in one direction, and positive charges can move in the other direction—it wasn't for nothing that electricity's basic property, electric current, is now measured in amperes.

He created a general rule based on the results of Ørsted's and his own experiments; the rule is called "Ampère's little guy," or more commonly "the right hand rule." We're not talking about one specific experiment; he suggested a rule to tell the direction of an electric current, of a magnetic field, or of motion (displacement), as long as two out of three directions are known. Electricity and magnetism were already connected in the minds of a number of scientists, but the scientific community couldn't wrap their minds around the idea that electricity and magnetism were one and the same thing; to them electricity simply caused effects on magnetism and the other way around.

Yes, the other way around. Ampère observed that, when a magnet moves through a copper coil, an electric current runs through the coil. With this experiment, he actually observed and discovered the phenomenon of electromagnetic induction, but he didn't follow up on that line of research. Because when it came to magnetism, he was more interested in the nature of magnetism. He was convinced that it was a microscopic form of dynamic electricity. He envisioned an electric current inside a magnet at the molecular level—that is, many microscopic electrically charged particles moving within the magnet. This idea was widely rejected for at least sixty years—right up until the discovery of the electron.

Ampère created a theorem, Ampère's law, which says: In a nearly static or stationary regime, in a vacuum, the magnetic field created by a current distribution has a circulation around a closed loop that is equal to the algebraic sum of the currents passing through the directed loop multiplied by the vacuum permeability.

We'll overlook the fact that this single sentence calls on a dozen concepts we cover in this book *and* it's completely abstruse. So what is this law actually saying? Well, I can tell you, this law describes the mathematical relationship between an electric current and the magnetic field generated by that electric current. There's a mathematical equation hiding behind that long sentence. It's the first of the Four Pillars of Electromagnetism; no one even knew what they were building up to yet.

Carl Friedrich Gauss, a German astronomer and physicist—and certainly one of the greatest mathematicians of all time—was also interested in electricity and magnetism. Perhaps it's more expedient to say that *everyone* in the scientific community was interested in electricity and magnetism. In 1831, the results of Gauss and Wilhelm Weber's work on magnetism contributed to Kirchhoff's laws for electricity.

Well, besides that, Gauss learned all kinds of stuff about magnetism and electricity. Amazing, powerful stuff! He himself installed one of the Four Pillars of Electromagnetism using one law of electricity.

Gauss's law on electricity is actually nothing other than Coulomb's law, but Gauss phrased it a little differently. Remember, Coulomb's law was about the fact that two charges of the same sign repel and two charges with opposite signs attract; it also stated how strongly these charges repelled and attracted, and provided a mathematical formula for calculating the force of attraction or repulsion. We can derive Coulomb's law from Gauss's law and vice versa. The only difference is that Gauss's law applies to an electric field that varies with time, which makes it a more general application of Coulomb's law.

This is what Gauss says—my apologies to the purists, but I want my readers to enjoy this book, so I'm going for maximum simplification here: Turning back to the analogy of a river from a little bit ago, if you had an electric river, you would say "electric flux" instead of "flow." Gauss said: The electric flux that passes through a sphere and that is produced by an electric charge located at the center of the sphere is proportional to the quantity of this electric charge. In other words, if the charge is bigger, the electric flow is bigger. This law can be extended to apply to a group of charges, even a group that includes charges with opposite signs. The calculation becomes a lot more complicated if the group doesn't have an axis or center of symmetry. But in classic cases—such as a straight wire approximated by a long thin cylinder—calculations are relatively straightforward for someone who has a good grasp of the right mathematical tools. The law also geometrically describes how electric fields form, going from the positive charges toward the negative charges.

Next up, Thomson. You remember Lord Kelvin? He showed us that a magnetic field was different from an electric field: An electric field can be generated by a single electric charge. Whereas a magnetic field doesn't exist like that; it's impossible to have a single-pole magnet. A magnet always has two poles—breaking it doesn't separate the poles; it only creates two new poles at the break. Magnetic field lines always go from one pole to the other. This is another fundamental difference between an electric field and a magnetic field; electric field lines can extend out to infinity. An electric field around an electric charge is often illustrated by a sea urchin; the charge is in the center and the spines are the field lines. In magnetism, that doesn't exist.

And finally, Michael Faraday, a British physicist and chemist and electrician . . . wait, electrician? When describing the preceding scientists, I could have said "electrician"; back in those days *electrician* was the word for a scientist who studied electricity. I didn't say they were electricians because I didn't want you to imagine Faraday or Gauss replacing a wall outlet at your house, on all-fours behind your sofa, displaying his workman's crack in all its glory—it can't be unseen! Anyway, Faraday doesn't score the final blow, but he piles up the ammunition for another scientist to blow the hole through the wall. Faraday is a *super saiyan*[33] of electricity, the Chuck Norris of electromagnetism, because Faraday discovered electromagnetic induction. And from there on, everything changed. Wait! Just a few pages ago, didn't we hear *Ampère* discovered electromagnetic induction? Yes, indeed, that's true. But I also said he didn't do anything with it. Ampère simply noted the finding and didn't look into it any further; he was pretty busy with other areas of study. But Faraday, he seriously got into it. He

[33] Name given to super warriors from planet Vegeta in the manga series *Dragon Ball* by Akira Toriyama (1984).

paved the way for being able to mathematically quantify, using a single equation, the variation of a magnetic field induced by an electric field.

Lorentz Force (or Abraham-Lorentz Force)

Hendrik Lorentz, a Dutch physicist, asked an interesting question. When an electric current runs through a wire and the wire moves around the magnet, the electric charges are moving through the conductor. We know where the force that moves the wire comes from. But when no current is running through the wire, the electric charges are at rest. What force puts the charges in motion when the magnet moves?

He went on to describe the force; we now call it the Lorentz force or Abraham-Lorentz force. But this force posed some problems that could be solved only by some really hairy physics.

In fact, though the equations for calculating this force completely agree with observation, it seems that the equations are completely false below a certain size of scale. Below this size, classical electromagnetic equations just don't work anymore. To explain this, without going too far into the technical details, let's say that when the scale is too small, the charged object would create an unlimited amount of energy by interacting with its own field—which is impossible. It would be like getting an object to accelerate before it's subjected to a force; it absolutely violates the causality principle.

This issue caused major problems for classical mechanics physicists to whom an electron—the only known particle at the time—was an ideal particle, meaning an infinitely small point. Now they had to describe this particle as having a radius, changing the point into a ball; so, they had to start talking about a *classical electron radius*. Below that radius, classical electromagnetism couldn't handle reality. Everyone would just have to wait for quantum mechanics to resolve the issue.

Making induction happen is easy enough: You get a wire coil—a wire wrapped around like a spring—and you place a bar magnet inside the coil. You connect the coil to a light bulb. Nothing happens. Good.

Now, move the bar magnet and "Let there be light!" Like I said, it's very simple to do. Except, that's not how Faraday did it the first time.

Faraday put the end of a wire in a mercury bath that had a permanent magnet in the middle of it. When he ran a current through the wire, the end of the wire started revolving around the magnet. It converted electricity into continuous circular motion! Ta-da! Faraday had just invented an electric motor. That was in 1821. And it was *ten years* later, on August 29, 1831, that Faraday discovered electromagnetic induction. Until then, they had always used a battery to make a current run through a wire. But here, a current was caused by moving a magnet. Mechanical motion of a magnetic field near a conductor converted the magnetic field into electric current. In other words, Faraday produced electricity by moving a magnet!!!!

Faraday, therefore, erected pillar number three of the Four Pillars of Electromagnetism. Construction isn't quite finished yet—but with Faraday's published work, we have all the raw materials needed. At this point, we hand the work to the master builder, he who would see farther than all of the others. That would be none other than James Clerk Maxwell.

22. Maxwell's Four Equations

James Clerk Maxwell, physicist, mathematician, electrician—oh my, that workman's crack, the image just won't go away—a Scotsman, and former student of Trinity College . . . you know the routine. If I made a list of all the physicists in history and selected the best one from each century, Maxwell would definitely be my pick for the physicist of the nineteenth century, along with Einstein for the physicist of the twentieth century. Maxwell studied electricity, magnetism, and induction, and he saw something no one before him had seen: a link. It seemed to him that something was missing, like a missing link. Maxwell dedicated his life to discovering this missing link, to bringing together electricity and magnetism; his desire was to simplify things enough to render the theory as simple and elegant as possible: Four simple yet powerful equations were the result of his work—actually, Oliver Heaviside wrote the current version of these equations, but it was really Maxwell who did the brunt of the work.

I realize this all seems a bit confusing. Actually, that's a pretty good description of the scientific situation in the late nineteenth century. Everyone was abuzz with the excitement of electricity and magnetism. Almost everyone was working on electricity and magnetism. Scientists felt the two subjects were at least related, but they just couldn't quite understand the nature of the relationship. Enter Maxwell.

First, Maxwell added to Ampère's equation, which then became the Maxwell-Ampère equation. Ampère had determined that an electric current generated a magnetic field, and he provided the equation for going from one to the other. Maxwell's addition to Ampère's equation showed that it's the same with a changing electric field; that seems like no big deal, but we just went from working with a static condition to working with a condition that changes with time.

Add that to the Faraday equation, which shows that moving a magnet generates an electric field, and then you can see that magnetic fields and electric fields are coupled—the change in one is proportional to the magnitude of the other. These two equations show that electric and magnetic fields act just like a swinging pendulum.

What these two equations say is that the magnetic component (bonus three-syllable word for "part") of an electromagnetic wave can be converted into an electric component and vice versa. Like the swing swapping energy between kinetic and potential forms, electromagnetic waves swap energy between electric and magnetic forms.

How's that again . . . electromagnetic waves? Where did that just come from?

We'll get there. But first, let's pick up two other equations, one from Gauss and one from Thomson (Lord Kelvin). The first one defines an electric field around one or more electric charges. The second one defines a magnetic field around a magnet.

At the start, Maxwell had *twenty* equations to work with. He spent several years whittling down the equations, finding some of them were derived naturally from others and some were the same as others, just

The Swinging Pendulum

Imagine a swing on the playground. Before starting to swing, you sit on the swing, then you back up, to get a boost; what you're really doing is getting higher. The seat of the swing (and your bum) are now higher than when at rest, just sitting there, quietly hanging from the swing set. If you let go, you know the swing will drop due to gravity. At the moment, though, you're not letting it do that. The swing and your bum have stored gravitational potential energy just begging to be set free. Now you lift your feet, and you start to move. Wheeeee! The gravitational potential energy is gradually converted to kinetic energy—that is, energy from motion. Once the swing passes its vertical axis, it continues to move because of momentum, but it starts to slow down as you and the swing begin to go up again. From that point on, you're converting kinetic energy into gravitational potential energy. Once the swing stops at its highest point, you'll begin to move in the opposite direction and the process begins again.

This pendulous motion—yes, because it's the same for a pendulum as for a swing, except you're not along for the ride—is an oscillating motion, swinging front to back. If there were no loss of energy (which is impossible), no air resistance, and no loss of energy in the form of heat, you'd have perpetual motion—the motion would continue forever.

written differently. Basically, he wanted to bring elegance to his theory of electricity and magnetism to create one unified theory of *electromagnetism*.[34] And though Heaviside was the one who ultimately wrote the four equations in their final form—Maxwell had already reduced the initial twenty equations down to eight—certainly Maxwell was primarily responsible for the advance. He succeeded in combining two phenomena that others thought were related only by causality. He showed they were indeed the same phenomenon in all cases—but then he went even further!

So, why am I talking about electromagnetic waves? Because electromagnetic waves are defined as variations—at least at the macroscopic scale—in electric and magnetic fields, sort of a wave on the surface of

---

[34]   And to those who can read them, the equations are just emanating elegance!

electromagnetic fields.[35] And it's possible to derive (nerd-word for "figure out") Maxwell's equations, because an electromagnetic wave is happy to propagate in an electromagnetic field. So, it doesn't need a medium—like sound waves need air as the medium to propagate. An electromagnetic wave can propagate in a vacuum. This was a crucial discovery—Einstein used it when he came up with the special theory of relativity. Maxwell used experimental data to calculate the speed of these electromagnetic waves. He found their speed to be approximately 193,085 miles/second. Hmmm, that number seems familiar . . . isn't 186,000 miles/second the speed of light? So it is! Thus in 1864, he suggested that light itself was actually an electromagnetic wave:

> The agreement of the results seems to show that light and magnetism are affections of the same substance, and that light is an electromagnetic disturbance propagated through the field according to electromagnetic laws.[36]

The Four Interactions

There are four types of interactions (sometimes called forces) in our universe—four frameworks that make it possible to describe the laws of physics: gravity, electromagnetism, weak nuclear interactions, and strong nuclear interactions. Gravity—it's counterintuitive, but gravity is the weakest of all the interactions. (Think about it: Hold your phone in your hand so it doesn't fall on the ground, and with the incredible strength of your muscles, you've just overcome the force of gravity . . . with your hand . . . I'm speechless— that's what describes how things with mass attract each other.) Regarding electromagnetism, we just had a whole chapter on it; no need for an epilogue here. As for the last two interactions, weak nuclear interactions relate to how radioactivity works, and strong nuclear interactions are concerned with how positively charged protons can stay side by side in atomic nuclei.

---

[35]   Not "sort of"—that's exactly it.

[36]   From "A Dynamical Theory of the Electromagnetic Field" (1865), p. 115.

Thus Maxwell united electromagnetism and optics. Let's think about this—it means other than interactions associated with gravity, electromagnetism describes *any and all* physical interactions at our scale!

Now, so far we learned that, at our scale, electromagnetism converts electric energy and magnetic energy into three forms.

Kinetic energy: creating motion with an electric motor.

Heat energy: creating heat by running a current through a resistant conductor.

Chemical energy: separating water into oxygen and hydrogen with electricity.

It's a lot easier now to understand how the compass was affected by the earth's magnetic field, the magnetosphere, which is a veritable shield protecting against constant emissions from the sun.

THE SOLAR SYSTEM

A unique place, like so many others.

System: n., group of items considered in their relationship to each other and functioning as a unified whole: *The nervous system. Various political systems.*

Solar: adj., having to do with the sun: *Solar radiation. Solar energy.*

The solar system is the group of all bodies subject to the gravitational influence of the sun, plus the sun—obviously. This includes the planets and their moons, asteroids, comets, and all the gases, dust, and debris we find strewn here and there throughout the solar system. But most important, in the middle of all that, of course, is the sun.

23. The Sun

When the ancient philosophers raised their eyes to the heavens, they saw innumerable celestial lights. If observed for only a few moments, these lights appeared to be standing still, but as the night progressed they moved from east to west across the sky. However, some followed a different path than the others: the sun, the moon, and five small lights, otherwise no different from the others. The Greeks called these wandering lights "planets."[37] We had to wait centuries for the hierarchy of space to be set to rights—or at least to get the sun, the moon, and Jupiter out of the same category. Today, we know the sun is not a planet; it's a star. But what *is* a star?

---

[37]   From the Greek πλανήτης, which means "wanderer."

A star is a single, compact, opaque mass of gas in hydrostatic equilibrium with a core that combines elements through thermonuclear fusion. Bon appétit!

Would you like some more tasty details? *Single mass*: This sounds stupid, but a star only exists in a single piece; even if two stars are right next to each other and exchange matter with each other, they are still considered two distinct stars. *Compact*: A star has a clearly defined shape, in contrast to, say, nebulae, which are gigantic, diffuse clouds of gas—among other things. *Opaque*: Electromagnetic radiation can't pass through it, and you can't see through it—not with infrared, not with ultraviolet, not with microwaves, etc. *Hydrostatic equilibrium*: I'll give you more details in just a bit, but it means that the star is massive enough to have gravity enough to hold its shape—in this case, a ball—and at the same time, something prevents the star from collapsing in on itself, which balances the gravity and puts the star into the state called hydrostatic equilibrium. *Nuclear fusion*: The star's internal pressure is so great that it literally forces particles to stick to each other to form atoms.

Hydrostatic Equilibrium

Hydrostatics is the branch of fluid mechanics that studies fluids—gas or liquids—that aren't moving, that are *static*. We talk about hydrostatic equilibrium when a pressure gradient (*there*'s a highfalutin phrase for "pressure varies with distance") counterbalances gravitational forces. Let's take the earth, for example. The pressure of the atmosphere is greater closer to the ground. This air pressure gradient keeps gravity from compressing Earth's atmosphere, and gravity prevents the atmosphere from floating off into space. Equilibrium.

In space, because there isn't any up or down or sideways, gravity acts the same in all directions. The result is when a heavenly body is in hydrostatic equilibrium, its gravity forms it into the shape of a ball, since the gravity is equal in all directions, and because a ball has the same shape, no matter what direction you look at it from.

To put it simply, our sun is a star *like the others*. There are smaller stars, and there are other much larger stars. All considered, our sun's mass is about average for a star. I'm not going to say any more about the mass of the sun, because the numbers get pretty grotesque at that scale. I'll just stick with telling you that the sun's diameter is more than one hundred times Earth's diameter.

You might think the surface of the sun is hot at 5,778 K[38] (around 5,500°C), but it's frigid compared to the sun's core at—hang on to your seat—*15 million* Kelvins or degrees Celsius (at these temperatures, there isn't really any difference). And the core's diameter is one-fourth the sun's diameter. Inside the core, the extremely high temperature is due to the immense pressure of the sun's own gravity. This pressure is so great that stellar nucleosynthesis is possible.

24. Stellar Nucleosynthesis

I *love* this title: "Stellar Nucleosynthesis." It sounds so epic . . . even though it only means "atom making." Stellar nucleosynthesis is literally the manufacture of atoms inside a star. In this section, you'll learn just how phenomenal Dmitri Mendeleev's instincts were when he said the lightest elements are the most abundant in the universe, with three notable exceptions: lithium, beryllium, and boron. And back when you heard about Mendeleev, there was also a Focus Frame about spectroscopy (see page 15).

Until the invention of spectroscopy, no one had any idea about how atoms were made or where they came from. Once spectroscopy was invented, astrophysicists used spectrometers to determine the chemical composition of the sun, if only to compare it with things they were already familiar with: the earth and meteorites. The sun is primarily made of hydrogen (92.1 percent), helium (7.8 percent), and (in decreasing order) oxygen, carbon, nitrogen, neon, iron, silicon, magnesium, and traces of other elements. We can clearly see that the lightest elements are the most abundant. Now, how do we explain that?

---

38   K for Kelvin. Remember Lord Kelvin? We're measuring absolute temperature here; the difference in temperature in degrees Celsius is only 273.15.

Let's ask George Gamow. He was a Russian, then later an American, astronomer and physicist. He left the Soviet Union for good when he went to the 1933 Solvay conference in Brussels; he pretended his wife was his secretary—but never mind, that's not what we're talking about here. In 1942, Gamow was the first person to suggest that all matter present in the universe was created when the Big Bang happened—we'll get to that, of course, but you must know what the Big Bang is . . . the creation of the universe as we know it, 13.8 billion years ago. Well, Gamow was involved in developing this theory. He thought all the various elements were created just after the Big Bang by aggregating (clumping) of neutrons followed by beta (β) decay—a kind of radioactive decay during which a neutron can decay into a proton and an electron. This falls under the category of weak nuclear interactions. But if atoms were created in this fashion, considering how fast the universe cooled after the Big Bang, nothing more complex than lithium could have been created. That shoots a hole in that theory.

Around the same time, in 1939, Hans Bethe, a German physicist and astrophysicist who later became American (he fled Germany in 1933) published an article titled "Energy Production in Stars," in which he analyzed the means by which hydrogen in a star could be converted into helium.

In the beginning, there is a mass of hydrogen in a star—a monumental mass of hydrogen, a mega-massive-mess o' hydrogen! The star contracts due to its own mass; that's gravity at work. And the contraction causes the pressure of the hydrogen to go up, so the temperature goes up. When the temperature exceeds ten million degrees, the hydrogen nuclei have enough energy to overcome Coulomb's barrier. Remember, Coulomb's law states two same-sign electric charges repel each other. This repulsion has a certain energy level; if this level is exceeded, same sign charges can be forced to stay together. That's exactly what is going on in the star at this point. By binding protons in pairs, a cycle of transformations occurs that ultimately produces helium, which has two protons and two neutrons in its nucleus. This fusion sends a phenomenal amount of energy outward from the star's core. This counteracts and balances the gravity. The star has reached its first equilibrium and has now earned star status.

So now, the star is converting hydrogen into helium, and producing an insane amount of energy, and radiating in all directions in space, and enjoying equilibrium. This is the present situation of our sun. By the way, if you're wondering where the hydrogen came from in the first place, it was made during the Big Bang.

After a long, long while—a few million to one hundred billion years, depending on the mass of the star, a short twelve billion years for our sun—the hydrogen starts to run out, and the energy from it fusing into helium isn't enough to maintain the balance against gravity. Gravity starts to get the upper hand. If the star is big enough—at least one-third the mass of the sun—it will contract again, causing a higher pressure on the helium within the star and, therefore, a higher temperature—I'm sure we've heard that somewhere before. At around one hundred million degrees, this time it is helium that has enough energy to be fused into heavier atoms. This starts another cycle of transformations, which gives us some of the heavier atoms: carbon and oxygen. Once again, the energy released by the fusion process balances out the gravity, and the star reaches its second equilibrium. For our sun, this would be the last equilibrium, because the sun isn't massive enough for what comes next.

And here we go again, when there's no longer enough helium to produce the energy needed to maintain equilibrium, the star contracts again under its own mass and compresses the carbon and oxygen until they reach a temperature on the order of 1 billion degrees. At this temperature, carbon fuses to produce sodium, neon, and magnesium during the star's new equilibrium—if a star is twenty-five times the mass of the sun, this equilibrium lasts for approximately two hundred years. Then it's neon's turn for fusion, at 1.2 billion degrees, as well as magnesium and oxygen, and smaller amounts of heavier atoms up to bismuth, polonium, and lead. This fourth equilibrium is much shorter than the earlier ones—only a year for a star with the mass of twenty-five suns. Next, at 2 billion degrees, fusion of oxygen takes place to produce silicon, phosphorus, and sulfur. In addition to releasing tremendous amounts of energy, these transformations produce the protons, neutrons, and other particles needed to form other elements like chlorine, potassium, argon, and calcium. For a star with the mass of twenty-five suns, this phase lasts about five months. The death of the star is near. Once equilibrium fails, there's another contraction; the temperature reaches 3 billion degrees, and the last fusion begins, silicon fusion. All the elements up to iron are going to be manufactured during this short period of a few hours on the last day of the star's life.

Iron is the most stable of all the elements. Iron doesn't fuse, at least not in the core like the lighter elements. Ironic isn't it?[39] Lighter elements create energy through fusion. Iron and the heavier elements take in energy during their fusion process to produce the heavier elements. So how and when do we get the heavier elements? . . . Hold on, we're almost there!

Now, it gets a little crazy. When there is no more fuel for fusion, the star collapses rapidly in on its iron core and it implodes! Gravity's pressure contracts the matter in the core until it has the same density as an atomic nucleus. Nothing can get near it. The least little particle that approaches the core bounces right off. As a shock wave sweeps outward from the center of the star, it rekindles a short period of fusion in the outer layers of what remains of the star. During this explosive period, all the elements heavier than iron are synthesized in an energy-sucking kind of fusion by rapidly combining protons and neutrons. With all the turmoil, the numerous collisions release enormous amounts of energy, causing another shock wave, and the star becomes a supernova. At this point, matter synthesized in the star over billions of years is thrown into space in every direction. The core is so dense that it compresses the atoms within the core so much that the electrons are squashed into the atomic nuclei, thus converting the protons into neutrons. Phenomenal radiation is emitted from the poles of what is now a neutron star, aka a pulsar. Its core can be as small as ten miles across, and it spins around its axis keeping perfect time, turning up to several thousand times per second!

[39]   Sorry.

Everything that is part of you, my dear reader, and the book you're reading, and the air you breathe, is all made of carbon, oxygen, iron, lead, magnesium, calcium, uranium, sulfur, cobalt, etc. All these elements are our sole legacy from whole generations of dead stars. We are all stardust. These elements are nothing more than the remains of whole generations of dead stars. Yes, we are all star guts.

25. Formation of the Solar System

In space, we find hydrogen in large quantities; we also find other heavier elements that are residue, the remains of stars that exploded. All these thingamajigs tranquilly drift through space, a cloud of gases, dust, and debris, until the cloud becomes large enough and dense enough to begin forming hydrogen gas, H2. This kind of cloud is called a *molecular cloud*. To our knowledge, approximately 4.5 billion years ago, a molecular cloud drifted serenely through our galaxy. It was large enough to span between 7,000 and 20,000 astronomical units—an astronomical unit is the average distance between Earth and the sun; exactly half the distance between the point where Earth is farthest from the sun and the point where Earth is nearest to the sun, which is 93 million miles[40]— The cloud's mass is just a little more than the mass of our sun, and it is about to become our solar system.

This cloud is large enough and dense enough to begin slowly, gently collapsing in on itself under the influence of its own gravity: just a tiny bit more mass somewhere around the middle and here we go! Even though all the particles in this cloud seem to move any which way they want, the entire cloud has an average overall motion. You could say the whole cloud "rotates" around itself. This is important. In space, the laws of mechanics tell us this rotational motion has no reason to stop, so long as no force comes along and changes it: In practice, we call that *conservation of angular momentum*.[41] That is the reason the solar system will be flat.

The cloud begins to consolidate as it continues to spin and swirl.

---

[40]   By definition, its exact value is 149,597,870,700 meters.
[41]   See Newtonian mechanics, page 104.

The cloud eventually becomes a solar nebula, a huge disk, like a huge pancake, or like those old vinyl 33 rpm LPs. The center will become more dense and—since pressure increases with the density—hotter. At the center of the solar nebula, we find a compact, burning mass, but its thermonuclear fusion reactor hasn't switched on yet. The sun is about to be born; about one hundred thousand years have passed. For about fifty million years, the sun grows as its gravity continues to suck in more matter, gas, and dust from the disk, until its internal pressure is finally high enough for hydrogen fusion to begin and start synthesizing helium. It reaches its first hydrostatic equilibrium: The sun is born!

From then on, throughout our sun's first equilibrium—which is still going on today—the matter around the sun revolves more serenely around it. As the matter revolves around the sun, various gases and dust randomly meet up and clump together, then de-clump, then re-clump as may be. Eventually, within the large, swirling disk around the sun, some clumps (accretion disks) develop in multiple places. If they are lucky enough to avoid destruction when they collide with other chunks, they eventually grow to be too large to be destroyed by an impact with a smaller rock. When a rock runs into a larger chunk, gravity captures the rock and the chunk gets fattened up. This process, called *accretion*, is how the various planets formed around our sun; but for the time being, they are large rocks at a minimum and *planetesimals*[42] at best.

We should note: Near the sun, temperatures are extremely hot, so no gases can condense or maintain solid form there. So, this close to the sun, within 4 astronomical units—four times the distance between the earth and the sun—only elements that evaporate at high temperatures are any good for use in forming planets—elements like iron and aluminum—which, by the way, are the rarest elements in space; this stunted the growth of the planets closest to the sun. Gases participate in forming the planets that are beyond 4 astronomical units from the sun. This virtual boundary sets apart the terrestrial planets—for now: Mercury, Venus, Earth, and Mars—and the gaseous planets—for now: Jupiter, Saturn, Uranus, and Neptune.

[42]   What a great word for tiny little planet.

I suppose you noticed how I specifically listed the planets "for now." It's not because I have any particular nostalgia for Pluto as a planet, but rather it's because when the terrestrial planets were formed, the sun had between fifty and one hundred planets for at least another hundred million years. These planets would crash into each other, and the largest terrestrial planets grew more. During collisions, debris was ejected. This debris grew more slowly, eventually becoming moons—this is no doubt how the Moon, with an uppercase *M*—our moon—was born. It's a little different with the gaseous planets: Scientists are still not quite sure how Uranus and Neptune could form at such a distance where there was so little gas available to build them. Scientists think perhaps they were formed closer to the sun, and they eventually migrated to their current orbit. This is called the *planet migration theory*, aka the Nice model.

The Nice Model

The model is named for Nice, a city located in southern France, the location of Côte d'Azur Observatory (Observatoire de la Côte d'Azur) where the model was initially developed. I won't go into details on orbital resonance and other mathematical complexities. I'll just simply say the Nice model is better at explaining phenomena such as late heavy bombardment, formation of the Oort cloud, the Kuiper asteroid belt, and the existence of the trojans of Jupiter and Neptune. These phenomena contradict earlier models about the solar system.

26. Mercury

It's not surprising that the closest planet to the sun experiences the most abuse from the sun. Mercury's volume and mass equal approximately 5 percent of Earth's. Its total surface area is 29 million square miles; that covers Asia and Africa. Mercury has no atmosphere—well, almost none anyway. In fact, any little bubble of gas on its surface is snatched away almost immediately by the solar winds. Mercury's "atmosphere" is so thin that the molecules almost never hit each other, and they provide about nil atmospheric pressure—the atmospheric pressure on Mercury is about two hundred billion times less than the air pressure on Earth at sea level.

Even though planet Mercury is close enough and large enough—distance-wise, anyway—to be visible from Earth with the naked eye, it is almost never seen. That's because it is always lined up so close to the Sun; the Sun's brightness overwhelms Mercury. I'm sure experts on the subject of planet Mercury have many wonderful and fascinating things to share about it, its formation, geology, position, etc. As far as I'm concerned, I just want point out a few interesting things about Mercury's orbit, because in the 1910s, Mercury's orbit was critically important to none other than Albert Einstein.

About Orbits

Although you are probably already familiar with the subject, I am sure it won't hurt to offer a few lines here about orbits, even if only to name a few characteristics. Traditionally—I would say "in classical mechanics," but we're not there yet—an orbit is a periodic, closed, curved path in space around a more massive object. For example, the moon is in orbit around the earth; the earth is in orbit around the sun.

This curved path is an ellipse, and the more massive object—actually, the combined center of gravity of the two objects, the *barycenter*—is one of the foci (plural of *focus*) of the ellipse. For the planets orbiting the sun, the sun is the barycenter. The point in the orbit farthest from the barycenter is called the *apoapsis*. For orbits around the earth, the *apoapsis* is called the *apogee*. For orbits around the sun, it's the *aphelion*. There are also specific words for the apoapsis of just about any celestial body, such as *apomelasma* for the farthest distance in an orbit around a black hole, or *apocytherion* for an orbit around Venus; we have three options for the apoapsis in an orbit around our moon: *apocynthian*, *aposelene*, and *apolune*. The point in the orbit closest to the barycenter is called the *periapsis*—and *peri-* works the same way: *perigee*, *perihelion*, *perimelasma*, etc. . . . you get it.

Mercury's perihelion—the point in its orbit closest to the sun—changes with each trip around the sun. This means that every time Mercury passes through the perihelion of its orbit, it moves a little. It's called the *precession* of Mercury's orbit. It has a periodicity of 225,000 years, meaning the orbit gets back to the same position every 225,000 years. The phenomenon was first observed in the nineteenth century, but no one could explain it. Shockingly, classical mechanics—Newtonian mechanics—which had explained the motion of the stars with such remarkable accuracy for nearly two hundred years, could not be used to explain this kind of movement!

I'll admit, not many people lost much sleep over this precession issue; the change is relatively small. The variation in Mercury's orbit is on the order of forty-two arcseconds per hundred years. This forty-two arcseconds is about equivalent to the height of this book seen from half a mile away. If you look at it that way: Mercury's motion varies between the top of the book to the bottom of the book . . . over the course of one hundred years. That is a *seriously* small amount.

So, it wasn't a big deal, because it didn't interfere with correctly measuring the movement of the moon, Jupiter, etc.

Arc Degrees, Minutes, and Seconds

From Earth, we use angles to define relative distances. This method involves trigonometry—I know a bunch of you are about ready to faint from just reading the word, so I won't get carried away with explanations. In a nutshell, it's good to know that a complete circle has 360°; a right angle, which is a quarter of a circle, has 90°, and a straight angle is worth 180°, etc.

Just like an hour of time is divided into sixty minutes, an arcdegree is divided evenly into sixty smaller angles, called arcminutes. And each arcminute is divided into sixty equal sections, called arcseconds. An arcsecond is, therefore, one sixtieth of one sixtieth of a degree and equals 0.000277°—oh, of course, what else would it be?! It takes 3,600 seconds to make a degree, same as the number of seconds in an hour.

But if you wanted to be rigorous, or if you were particularly interested in studying Mercury's motion, it was indeed a problem.

The French astronomer Urbain Le Verrier tried to shed some light on the mystery. He used classical mechanics to calculate Mercury's movement around the sun and accounted for the influence of all the then-known planets in the solar system. He thought the influence from large planets like Jupiter might be significant enough to explain the variation in Mercury's orbit. In September 1859, he presented his results, which disagreed with observations. Then in 1882, Simon Newcomb, an American astronomer, refined the results by taking into account the fact that the sun rotates on its own axis (this slightly flattens the sun at the poles and also slightly increases its diameter at the equator), but to no avail. It still didn't match the observations.

In 1860, Le Verrier tried a new approach inspired by the recent discovery of Neptune. Discovered in 1846, Neptune is the result of some excellent brainwork on the part of some astronomers. They thought disturbances in Uranus's orbit, which they didn't understand, were due to the existence of a more distant planet, which they didn't know about—more details are in chapter 39, "Uranus and Neptune" (beginning at page 140).

The discovery inspired Le Verrier to go looking for an undiscovered planet between the sun and Mercury. He even named it Vulcan . . . ahead of time. He generated a huge pile of calculations to determine the exact location of said planet, then the astronomers pointed their telescopes in that direction and found . . . nothing. Vulcan didn't exist. Everyone would have to wait for Albert Einstein's theory of general relativity before they could understand why Mercury's perihelion advanced over time.

27. Good Old Science Dude: Guillaume Le Gentil—Gentle Willy

Venus is the second-closest planet to the sun—yes, I am aware that this section on Guillaume Le Gentil appears right smack in the middle of a chapter about the solar system . . . relax, I have a plan.

Venus is a peculiar planet. It is immediately identifiable to the naked eye at dawn and dusk. Venus has several nicknames: the shepherd's star,

the morning star, and the evening star. Venus is the second-brightest
celestial body in the sky, after the sun and the moon. Surely, its brilliance[43] inspired naming it after the ancient Roman goddess of beauty;
very poetic for a planet covered with a thick, toxic cloud of carbon dioxide and sulfuric acid. Venus is peculiar for several reasons. First of all,
it's the only planet, other than Uranus, that rotates in retrograde, meaning it rotates opposite from the way it revolves around the sun. It's also
the only planet in the solar system that has a day—one complete turn on
its axis—that is longer than its year—one complete trip around the sun.
Venus is just about the same size as the earth, yet Venus brags that her
waistline is 5 percent smaller than Earth's. Since Venus rotates so slowly,
there's not much of a bulge at the equator. It's the most spherical body—
she has the best curves—in the solar system. In addition, its surface is
relatively young—even her skin is more youthful!—possibly because
the surface is sporadically rejuvenated by intense volcanic activity. Also
peculiar, the volcanoes don't seem to put out lava flows; this is unique in
the solar system and isn't well understood even today. Last, its orbit is
almost a perfect circle.[44]

Venus is nearly the same size as the earth, as I mentioned, but its
atmosphere is nearly one hundred times more massive than Earth's. It
has the thickest atmosphere of all the terrestrial planets. The atmosphere has a unique dynamic; it turns in the opposite direction of
Venus's rotation. The speed helps the atmosphere make a complete trip
around the planet in less than one hundred hours. The highest winds in
the atmosphere have been clocked at a ground speed of more than 220
mph. Since the atmosphere is 96 percent carbon dioxide—so dense, it's
nearly in liquid form in the atmosphere—Venus suffers from an extreme
greenhouse effect. This causes temperatures on Venus to be even hotter
than those on Mercury. Mercury has no significant atmosphere and is
twice as close to the sun—temperatures on Venus are around 860°F.
Venus is literally the solar system's own hell planet—I specifically say
"planet" because there are a couple of hell moons that are pretty cool.

[43]   Its *albedo*, actually; its ability to *reflect* light.
[44]   Other planets have elliptical orbits that are more eccentric.

In the history of astronomy, Venus was probably immediately interesting to scientists, because after the sun and moon, it's the most easily identifiable celestial object. Venus also made it possible to accurately measure the distance between the sun and the earth in 1769. Astronomers used the *parallax* method described by James Gregory, a Scottish astronomer and mathematician, in 1663 . . . some 106 years earlier.

Parallax

Technically, parallax is the change in the angle at which an observer sees an object when it moves—"it" can refer to either the object or the observer.

In practice, parallax is what happens when you are sitting in a train next to a window. When you look out the window, the scenery that's farthest away seems to move by more slowly than the flora and fauna right next to the train. It's a phenomenon you notice when driving at night. It's what makes the moon seem to follow you.

Astronomers can compare the movement of two separate objects in the sky and accurately determine how far away they are from the observation tool, usually a telescope. And it's the method we're going to apply to Venus.

There is a beautiful phenomenon known as the *transit of Venus*. Every so often, Venus crosses directly between the earth and the sun. From the earth's perspective, Venus partially eclipses the sun—*seriously partially*; Venus appears thirty times smaller than the sun. Using a suitable telescope, one that protects you from being blinded by the sun, you can see a small black disk crossing the width of the sun. This black disk is Venus silhouetted against the sun. By looking at the transit of Venus from multiple locations on Earth, astronomers can use the parallax method to accurately determine the distance between the sun and the earth at the time of the measurement.

Using the parallax method, Edmund Halley, a British astronomer, developed a protocol for observing the transit of Venus. Yes, *the* Edmund Halley known for the famous comet bearing his name—he didn't discover the comet, but he did determine its periodicity (how often it comes around), so when it reappeared sixteen years after his death, as he predicted it would, the comet was named in his honor. And only twenty years after Halley's death, the protocol was used by Mikhail Lomonosov, a brilliant Russian physicist, astronomer, historian, poet, and more. He organized a joint international effort for the observation of the next transit of Venus. This involved a hundred different astronomers located here and there, all over the world.

Among these one hundred astronomers, there was a very promising Frenchman, Guillaume Joseph Hyacinthe Jean-Baptiste Le Gentil de la Galaisière, or simply Guillaume Le Gentil in French. In the interest of simplicity, and to make it easier for you to identify with him, and with a marvelous stroke of overfamiliarity, from now on we will call him Gentle Willy. Gentle Willy's destiny is going to be quite out of the ordinary. He is the Inspector Clouseau of astronomy, the Homer Simpson of science. Beyond a doubt, he is one of the most *epic* losers in the long and tragic history of loserdom. Thus I plan to shout FAIL! at those points in his life when it seems as though karma were making him pay back *all* of his mistakes, from *all* of his previous lives, over the course of a few years—because, yes, karma is a bitch.

Gentle Willy, who lived in Paris, chose Pondicherry, India, as the location for his observation. At the time, it was the seat of government of the French East India Company—a French colony, if you will. Pondicherry is located way over on the far east side of India. Since the Suez Canal hadn't been built yet, to get to Pondicherry by boat, Gentle Willy had to take a ship all the way around the southern tip of Africa. Gentle Willy planned to observe the transit of Venus expected in June 1761, so he left in March 1760, giving himself sixteen months to get there. That was certainly more than enough time. He hugged his wife and probably said, "See you soon," and "I'll be sure to write," completely unaware of the destiny that awaited him once he left Paris.

Four months later, in July, he had already gotten as far as Isle de France (Island of France), way over past Madagascar—it is called Mauritius Island now, but at the time it was called Isle de France—so far, so good. He stayed there until March 1761, preparing for the next leg of his odyssey, planning how he would observe the transit of Venus, what measurements he would make, how he would take them, etc. He wrote a letter to his wife and sent another to the French Royal Academy of Sciences. Somehow, for reasons unknown—shipwreck, pirates, or whatever—neither letter reached its destination—FAIL!

In March 1761, he embarked for Pondicherry aboard a fast frigate, *La Sylphide*. As they neared the end of the voyage, the crew learned that the war between France and England, the Seven Years' War, had reached Pondicherry. After several months of siege, the city had fallen on January 15. Pondicherry was no longer a safe place for a Frenchman, though he was only an astronomer, one sent by order of the king of France to boot. The ship made a U-turn—FAIL!

Gentle Willy hoped to be able to return safely to Isle de France to make his observations there. At least, it would be better than nothing. When Venus crossed the disk of the sun, on June 6, 1761, the skies were clear and ideal for observations; unfortunately, *La Sylphide* was still at sea. Gentle Willy tried to make his measurements as accurately as possible, but it was simply impossible on the moving ship with the swell of the sea. After a sixteen-month voyage to the end of the earth, Gentle Willy had to face the obvious: complete and utter failure of his mission—FAIL!

Let's have a closer look at the periodicity of the transit of Venus. It's a rare phenomenon that occurs approximately twice every 150 years. But it always happens in "pairs" of transits; you wait 100 years or so and you get a transit, then you get a second transit 8 years later. After that, you have to wait more than 100 years to observe the phenomenon again. The transit of Venus on June 6, 1761, was the first of two transits; the second one was due to occur June 3, 1769—the next one after that would be December 9, 1874. Well, Gentle Willy decided to stay where he was and wait for the next transit. So, he wrote a letter to his wife and another to the French Royal Academy of Sciences to inform them his trip would require an extra 8 years. You guessed it. Again, the letters never arrived at their destinations—FAIL!

To kill time, he studied the east coast of Madagascar, gathering information to improve the maps of that area. After a few months, he headed for Manila, his newly selected destination for making his observations of the transit of Venus—as if Madagascar weren't far enough from home! For your information, Manila is the present capital of the Philippines. It was founded in the sixteenth century by conquistadors. It was a thoroughly Spanish territory at the time. When Gentle Willy arrived in Manila, even though France and Spain had been allies in the Seven Years' War, they had a number of competing interests in Europe and the Americas, so there was an air of tension and suspicion. The Spanish government in Manila didn't trust Gentle Willy, sent by the French Royal Academy of Sciences and therefore by order of King Louis XV. They suspected he was a spy and it was politely suggested that he shove off as soon as possible—FAIL!

By then it was 1768 and the Seven Years' War was over. France had regained control of Pondicherry, and Gentle Willy could come ashore in peace. It seemed like fate was finally done riding his case. Gentle Willy wrote a letter to his wife and another letter to the academy to inform them that he was making preparations to observe the next transit of Venus. Neither of these letters reached their destination—FAIL!

Gentle Willy wanted to make absolutely sure he wouldn't miss his last shot—remember, the next transit of Venus wouldn't be until 1874, and Gentle Willy would be long gone by then. He decided to build an actual observatory at Pondicherry; simply putting a telescope on a rooftop just wouldn't do. He planned to make a perfect measurement for the glory and honor of both the French Royal Academy of Sciences and the king himself, while supporting the international scientific community in accurately determining the distance between the sun and the earth.

The observatory was built on the ruins of the old fort in Pondicherry. Gentle Willy began preparations by establishing the exact latitude and longitude of his observation. Everything was finally ready. The transit was to take place on June 3, 1769, and would last around six and a half hours. During the entire month of May, the weather was particularly clear. The observations Gentle Willy made while preparing for June 3 were fantastic. Local dignitaries were congratulating him in advance on the exceptional measurements he would make. On June 3, when the transit of Venus was about to begin, a cloud quietly floated over Pondicherry and covered up the sun! It stayed there until half an hour after the end of the transit, then the sky became exceptionally clear again—FAIL! Gentle Willy didn't get a chance to record a single observation. Not one. To add insult to injury, we should note that the weather in Manila was absolutely magnificent for the entire transit—FAIL!

The story could end there, and that is the case as it relates to Venus, but the truth is that fortune hadn't finished smiling on Gentle Willy, and this smile was the smile of a hungry lion that just noticed a fat zebra with a broken leg—that is, if lions could smile.

As you'd imagine, after all these misadventures, Gentle Willy was unhappy; he sank into depression. It had been nine years since he'd seen his wife. After several months, he decided to go back to France, but his departure was delayed by a particularly bad case of dysentery—FAIL! After failing to die from dysentery, he finally left Pondicherry in March 1770, heading for Isle de France, where he got sick again—FAIL! And he was stuck on Isle de France for seven months more. He couldn't stand Isle de France any more. He only had one thing on his mind: getting back to Paris. He finally managed to leave, but only two weeks into the voyage, his ship sailed into a storm and they were forced to return to Isle de France—FAIL! He had to wait until March 1771 for a passing Spanish ship to agree to take him back to Europe. They dropped him off at Cadiz, Spain; he made his way across Spain and eventually arrived in Paris in October 1771 . . . more than eleven years after his departure.

Once in Paris, he discovered a terrible truth: Several years earlier, he had been declared *dead*—FAIL! His wife had remarried—FAIL! His vacant position at the academy had been filled by someone else—FAIL! His estate was in the process of being divided among his heirs—COMBO FAIL!

He had his lawyer running things while he was gone. When Gentle Willy asked him for the money that was left, the guy took the money and ran—FAIL! Gentle Willy then sued his heirs, requesting that his estate be returned. And. He. Loses. What an incredibly epic, magnificent—FAIL! Let's take a moment and let this sink in: He actually lost the lawsuit to get his estate back, even though he *wasn't actually dead*! The king himself had to intervene to get Gentle Willy another position at the French Royal Academy of Sciences. In the end, after many costly legal actions, he regained what was rightly his from his heirs.

He met a rich heiress with whom he fell desperately in love, and they lived happily ever after until Gentle Willy's death in 1792. Amazingly, despite the astronomical length of his full name, indicating he was a noble and therefore a likely invitee to the guillotine, Guillaume Joseph Hyacinthe Jean-Baptiste Le Gentil de la Galaisière was not harmed during the turmoil of the French Revolution.

Here, at the conclusion of this incredible story about the life of our Gentle Willy—Guillaume Le Gentil—let's take a moment to honor all the scientists who never found anything. Everyone has heard of Albert Einstein, Marie Curie, Isaac Newton, Galileo, and Copernicus, even if they don't know what these luminaries accomplished. But for every Albert Einstein, there are hundreds, thousands even, of scientists who dedicate their *entire lives* to research, study, and understanding of our world. And it's a good thing to take a moment, now and then, to think of the scientists whose seemingly fruitless work often provides something useful for later significant discoveries.

28. Earth, the Goldilocks of the Solar System

If you take home only one thing about planet Earth, even before we start talking about its oceans, its atmosphere, its temperature, or its moon, I want you to remember—as far as we know—Earth is the only planet in the solar system that supports life. Life as we know it, anyway. Make no mistake here. In this paragraph, the key phrases are *as far as we know* and *as we know it*. It takes a certain kind of (overly confident) person to assume there can be no form of life other than carbon-based life that requires liquid water, and that there is no other life beneath the surface of the sun, say—it may seem improbable, but improbable is not proof. We have some civilizations that think that a rock is life, so why can't we accept that, according to their definition, there is life everywhere in the universe, including in space itself?

I know you're thinking: What the hell is this guy talking about? Or as those with a more proper upbringing might say: This gentleman's musings utterly elude me. Yes, I am completely aware that in this, the section called "Earth, the Goldilocks of the Solar System," I am instead talking about living rocks distributed throughout the universe. But believe me, I know where I'm going with this . . . pretty much, anyway . . . because the critical thing on Earth is . . . Us. Not "us" meaning people—the species, not the magazine—no! *Us* meaning the living: humans, certainly, but also all mammals, all animals as well as plants, bacteria, fungi, earthworms, even mosquitoes! Yes, I did. I said "even mosquitoes." And I'd say it again, in spite of the universal dislike for these insects, alone responsible for nearly half of all human mortality on Earth since humans arrived on the scene.[45]

For a few moments, let's focus on "life as we know it," and let's try to understand what makes life possible. Life on Earth is based on an abundance of organic compounds composed of nitrogen, oxygen, hydrogen, and carbon. The presence of these elements is necessary for life on Earth, but that's not the only thing needed. For the various chemical reactions that are definitely necessary for life to emerge on Earth, a heat source is required; one that is not too strong, not too weak—it has to be just right, just enough to keep it warm. But that's not all. There has to be a propitious[46] environment for life to emerge, with liquid water that's not too cold—not ice—and not too hot—not steam. The story of life on Earth, without going into further details at this stage of the book, is the story of Goldilocks and the three bears, the children's story in which little Goldilocks goes into the three bears' house and finds three chairs: one is too big, one is too small, and one is "just right." She sits on it and settles in to eat from one of three bowls of porridge. One is too hot, one is too cold, and one is "just right." Then she finally falls asleep in the one bed out of three that is "just right." The appearance of life on Earth is

---

[45]   Mosquitoes carry malaria, an illness responsible for approximately 50 percent of all human deaths in the history of humanity—according to the American Association for the Advancement of Science.

[46]   *Propitious*, really? Why not just say favorable?

like a Goldilocks who made billions of "just right" decisions, without which life as we know it would not exist on Earth.

29. The Earth Is Round

The earth is round. Big deal. I haven't taught you anything new there. So how about I teach you some human understanding of the subject? There is a widespread idea that people—the species, not the magazine—believed, until rather recently, that the earth was flat. Well, that's incorrect. Humankind has believed in a round Earth since antiquity. Take the marble statue *Farnese Atlas* for instance. This statue from the second century, a Roman copy of a Greek statue, is the oldest existing statue showing the titan Atlas carrying "the celestial vault." The statue doesn't show Atlas carrying a plate in his hands; he's carrying a ball. The *celestial vault*, it's true, but a ball at whose center we find the earth.

How did humans figure out that the earth was round? At first it was simply an assumption. For them, the earth could just as easily have been a flat plate with a domed lid. So, how did they come to think of a ball-shaped Earth when observations never seemed to hint at the curvature of the earth? We often think of Pythagoras as first to think the earth was spherical, but Parmenides of Elea, a contemporary of Pythagoras, taught that the earth was spherical and isolated in space. He supported the idea simply because "there's no reason to fall on one side or the other."[47] During that same epoch—in the fifth century BCE—Anaxagoras of Clazomenae said that the moon was not just a changing disk but a sphere, and he put forth a completely correct theory on the eclipses—you can definitely tell that Anaxagoras of Clazomenae is one of those guys with great instincts about the world around him (see pages 11 and 12). This brings us to Aristotle, who, for once, was brilliant.

Aristotle deduced that the earth was spherical by the round edge of the shadow of the earth during an eclipse; that's one thing that he definitely gets credit for. This made it possible for him to conclude—how? nobody really knows—that the circumference of the earth is 400,000

---

[47]   Remarkable observation for the era.

stadia,[48] somewhere between 39,000 and 46,000 miles, which is absolutely wrong, but nice try. One of the main indications of a spherical Earth, or at least the curved nature of the earth's surface, came from an observation from Alexandria, Egypt: As a boat got farther away from port, the hull disappeared below the horizon before the top of the mast. This meant, in the distance, the surface of the sea was lower than the horizon. In Alexandria, there was also a librarian named Eratosthenes—not just some guy working at the library around the corner, but actually the head librarian at the Great Library of Alexandria. Eratosthenes developed a more accurate method of measuring the circumference of the earth. Other than his mind, all he needed was a stick and . . . a camel.

30. Good Old Science Dude: Eratosthenes

Between 230 and 193 BCE, Eratosthenes was the head librarian, or director, of the Library of Alexandria, the largest library of ancient times. This library held more than seven hundred thousand volumes during its greatest era, which was during the reign of Julius Caesar. Everything ever written on philosophy, music, geometry, astronomy, poetry, theater . . . everything, or nearly anyway, was available at the Library of Alexandria.

As I said, in Eratosthenes's time, the idea that the earth was a sphere was widely accepted by scholars; Plato, Aristotle, and many others considered it as an accepted fact. But actually, no one knew for sure, so it was still a hypothesis. Eratosthenes knew geometry, so he figured he could accurately measure the earth's circumference. That he even started the effort shows exceptional intuition for that era. He thought that if the sun was sufficiently far away, he could assume the sun's rays were parallel when they hit the earth. That probably doesn't seem like such a big deal—I'm sure you don't see the relationship between that

---

[48] A stade is an ancient unit of measurement of a length no one is 100 percent sure about. The distance of a stadium's track in Greek footraces, the distance traveled in a minute of marching, or two minutes. . . . We have to guess what Aristotle meant, hence the range in miles.

and the circumference of the earth—but it's this intuitive step that made it possible for him to take real-life actions to find the answer.

Eratosthenes was very familiar with the city of Syene,[49] even though he lived in Alexandria.[50] He remembered there was a special well in Syene. On the summer solstice, June 21, when the sun was at the highest point in the sky, the sun's light didn't cast any shadow at the bottom of the well. He concluded that, at that precise moment, the sun was exactly vertical over the well. He also knew that at the same moment at the same time of year, the sun cast shadows all over the place in Alexandria, indicating that the sun wasn't at vertical anywhere in town. Because of this crazy insight—about the sun's rays being parallel—on summer solstice, in Alexandria, when the sun was at its highest point in the sky, Eratosthenes stuck a gnomon—yeah, that's a real word—vertically into the ground.

Gnomon

A gnomon is a perfectly straight stick, with a fixed, known length. It was used to take measurements as well as to communicate measurements. It worked great, as long as the people you were communicating with were using a gnomon of the same size.

He asked a friend, a follower—OK, somebody—to record the length of the shadow the gnomon cast on the ground at the specified time. Once the measurement was made, Eratosthenes followed this thought process: Since the sun's rays are parallel, the rays of light hitting the well at Syene are directly overhead—that is, vertical; therefore, they hit the earth's surface perpendicularly—that is at a right angle. The earth being round in theory, it follows that the rays, if extended, would continue to the center of the earth. But on the other hand, we know the sun's rays striking the gnomon don't hit it vertically, because it casts a shadow. Since the gnomon is placed vertically with respect to the ground, it's pointing straight toward the center of the earth. Consequently, this configuration creates congruent alternate

---

[49] Syene is the present-day Aswan.
[50] Alexandria, the present-day . . . Alexandria.

interior angles. So he could see that the angle created by the gnomon and the sun's rays hitting its tip would be the same as the angle between the sun's rays hitting the well and the line formed by the gnomon.

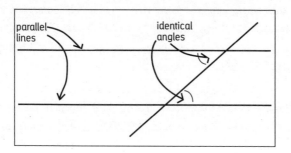

Alternate interior angles

Eratosthenes concluded: If he knew how many times to multiply this angle to make a complete circle, and if he knew the distance between Alexandria and Syene, all he had to do is multiply the distance by the number the angle was multiplied by, and the total would be a complete trip around the earth. QED![51]

The catch was—there was a "slight" catch—the trip from Alexandria to Syene was a hell of a hike. Eratosthenes had to call upon some distance professionals, the famous Egyptian surveyors, the bematists.

The bematists did their job, and gave Eratosthenes their estimate: The distance between Alexandria and Syene was 5,000 stadia, or approximately 575 miles. Then, he drew it out in two dimensions. He drew lines to represent the gnomon and the shadow to find the angle of incidence of the sun's rays, the angle at which the sun's rays hit the gnomon and the earth in Alexandria at the summer solstice. He figured out that to make a complete circle, the angle would have to be multiplied exactly fifty times—the measured angle was 7.2°; they didn't use degrees at the time, but anyway—he had to multiply the angle by fifty to get the 360° of a complete circle.

---

[51]   *Quod erat demonstrandum*, or "that which was to be demonstrated."

Bematist

A bematist was a surveyor in ancient Greece. The job consisted of traveling a given distance on the back of a camel and estimating the length of the distance by counting the steps of the camel—one step is one *bema*, or one pace. Yes, you read that right, by counting the camel's footsteps. Because, it seems, the steps of a camel are extremely regular. Wow, counting the steps of your camel going for hundreds of miles is a job that's . . . um . . . original.

Eratosthenes had his answer: the circumference of the earth is fifty times 5,000 stadias, which equals 250,000 stadias, or approximately 28,700 miles.[52]

Well? How close was he? Today, we know that due to its rotation the earth isn't a perfect ball; it's slightly flattened at the poles and it bulges slightly at the equator. If you measure the circumference of the earth by measuring around the equator versus around one of the meridians (a lovely word for "longitudinal lines")—a line that passes through the poles—you don't get the same result. On a map you can see the line between Alexandria and Syene is more like a meridian. With our modern measuring devices, we've determined the average length of a meridian, the earth's circumference measured by passing through the poles: the actual measurement is 24,900 miles. So Eratosthenes's answer was off by about 15 percent, which isn't too shabby for that time and with *that* technology—a camel for crying out loud!

That being said, we might want to moderate our reaction a bit, because the result is a culmination of numerous approximations, some of which could alter the final result in a favorable manner. Let's have a look: Determining the exact moment when the sun is at its zenith on the day of the summer solstice—that's relatively easy; you simply measure the shadow of the gnomon frequently and use the shortest shadow. Measuring the length of the shadow: That's the first approximation. Determining the angle: That's the second approximation. Finding the

---

[52]   The most widely accepted value for the length of the stade that Eratosthenes was using is the Italian stade, about a tenth of a mile.

number of times a complete circle contains this angle: That's the third approximation. Determining the distance from Alexandria to Syene on camelback, while not traveling in a straight line: That's yet another approximation—the biggest one. But still, the result is exceptional, and it's not surprising that it has stood through the ages. Speaking of age, a question just came to me—standard formula for an easy segue;[53] I think you know the question was planned way in advance, and it's just a clever trick to give you the impression that this is a real-time conversation, and this comment interrupted the book for no good reason to . . . uh, sorry . . . moving on—so, a question just occurred to me about the earth. We know its shape, we know its size, we know something about its atmosphere, and we know that oceans cover 70 percent of its surface, but how old is the earth?

31. How Old Is the Earth?

The earth is 4.54 billion years old, plus or minus 1 percent. There you go! I could let it go at that, but I think you know by now that most of the time, the important thing is not so much the answer as the path to finding it. Determining the age of the earth was quite an obstacle course, particularly for Clair Cameron ("Pat") Patterson, a twentieth-century American geochemist—yep, we had to wait until 1955 to obtain a relatively accurate age the earth and until 2012 for anything more accurate than that. There was no easy answer. There's one particular reason for that: the tectonic plates—I know that's plural, but I mean all of them—are the one reason. But we're getting ahead of ourselves; let's start from the beginning.

Aristotle thought that the earth had always existed; according to him, the earth was the center of the universe, and it wasn't completely out of line to think that this center of everything had always been there. On the other hand, some myths and religions offered the idea of a creation of

---

53 Pronounced "seg-way" but meaning "transition," not that two-wheeled motor-
   ized thing with handles that was supposed to revolution personal transportation
   but only made the pogo stick the second-most ridiculous transportation device.

the world, sometimes nearly instantaneously. For those who understand the Old Testament as a historic document to be taken literally, the earth should date back to approximately 4000 BCE. Many were the scholars who used the sacred scriptures to determine the exact date of creation. It should be noted that the Old Testament is fairly detailed regarding the chronology of the first few millennia: seven days for creation of the earth, then Adam, at the age of 130, fathering Seth, who was 105 years old when he fathered Enoch, who fathered . . . etc. Noah's birth is easily dated at 1,056 years after creation. The Flood arrived 600 years later and Abraham was born 292 years after that. So, the earth was allegedly 1,948 years old by then. We can continue, with a few hazy areas, until we catch up with history as recorded by historians, with Nebuchadnezzar II who destroyed the first temple in Jerusalem in 586 BCE.

Right up until the seventeenth century, it was uncommon for any-one, even scientists, to doubt that the Bible was true. And many were the scientists, including some of the most famous, who tried to reconcile their knowledge of astronomy and physics with the Bible—dating the Flood, finding the star of Bethlehem, providing a scientific explanation for the parting of the Red Sea, etc. It's why Kepler suggested the age of the earth was 3,993 years and why Newton suggested 3,998 years. To this day many are still trying to reconcile the Bible and science. Back then, it wasn't until Descartes that an idea—a wild idea then, even though it's commonly accepted today—began to gain acceptance: Today's laws of physics are universal and immutable in time; therefore, today's physical laws apply to yesterday and will apply to tomorrow. Descartes thought that God created the earth, of course, but then made it subject over time to the effects of the laws of physics—erosion, for example. Blaise Pascal complained about the idea, whining that Descartes wanted to have God, but only when it was convenient.[54]

As the seventeenth century rolled into the eighteenth century, a number of theories developed to determine the age of the earth. For example—and this took decades to figure out—by studying the strata, aka sedimentary rock layers, which are deposited very slowly;

[54]   "Since then, he has no more use for God" (Pensées).

concomitance (wowsa! what a word for "happened at the same time"!); and anteriority (what happened first) of some events compared to others. It's a little like looking at the life of a tree by examining the rings in its trunk, but there's a difference between counting layers of the earth and rings in a tree. We can figure out the life cycle and the period of time covered by each ring, but at the time, it was extremely difficult to figure out the timespan for a stratum of rock. It became evident that the sedimentary layers couldn't possibly have been formed over the course of a few thousand years; it was more like hundreds of thousands of years, perhaps even millions of years.

Benoît de Maillet, the French consul in Egypt, started with the idea that water covered all the land on the earth and then the water receded slowly. By estimating the rate at which the water receded, he concluded the earth was two billion years old—but, sigh, you know the story about avoiding problems with the still-powerful church of that time. For one thing, he made sure his works would not be published until after he died; second, they were to be published only in Holland, where the church had no control; on top of that, they were to be published under the pseudonym Telliamed, which is "de Maillet" spelled backward—this guy was obviously much better at studying the history of the earth than he was at cryptology. I know twelve-year-old kids who use this system to "guarantee" their anonymity on the internet.

Georges-Louis Leclerc, Comte (Count) de Buffon is up next. Inspired by de Maillet, he studied consolidation of the sedimentary layers in the Alps and the rate of deposition—formation of a layer of sediment—at the bottom of a sea. Buffon came to the conclusion that the earth could be at least seventy-five thousand years old. According to some sources, he may have been leaning toward an age as old as three billion years, but felt it wouldn't be "prudent" to publish those findings.[55]

In short, looking at layers of sediment—and at fossils—made it possible to piece events together relative to each other, but there was no way to establish an absolute age for the earth. Other efforts helped

[55]   Hubert Krivine, La Terre, des mythes au savoir (The Earth: From Myths to Knowledge), 2011.

advance the science: projects on studying changes in ocean salinity, and on studying variations in the distance between the earth and the moon. However, we had to wait for Pat Patterson, in the 1950s, to provide a more accurate determination of the earth's age. Patterson was an American geochemist with a degree in molecular spectroscopy. During World War II, he and his wife—also a scientist—worked on the now famous, but then top secret, Manhattan Project.[56] That's where Patterson was introduced to mass spectrometry—a field of study that makes it possible to accurately determine the isotopic composition[57]—whuuuut!?—of a given material. After the war, he returned to his studies and worked on his doctorate under the direction of his adviser, Harrison Brown, an American geochemist and nuclear chemist. Brown asked Pat to develop a method to date the earth—easy-peasy, right?

The biggest problem Patterson ran into was . . . well . . . er . . . the earth itself. That's because the earth's volcanic activity and the movement of tectonic plates were constantly rearranging the earth's rocky layer. Dating rocks, even the oldest ones, even the actor,[58] using the methods known at the time didn't guarantee that a particular rock dated all the way back to the beginning of the earth. Fortunately, with the knowledge gained from his study of radioactivity, Patterson realized it was possible to use mass spectrometry to determine the isotopic composition of meteorites. Patterson studied the isotopic composition of lead in meteorites—only in those meteorites he was convinced dated back to the creation of the solar system—and in 1955, he determined that the earth and the meteorites were the same age. They were created 4.55 billion years ago from the same initial source of material.

Just so you know: My attempt to simplify Patterson's accomplishment is an absolute insult to the exceptional quality of his work. It took years and years of analyzing rock after rock, sample after sample! It took

---

[56]   The Manhattan Project was a secret American program active during World War II to build the first atomic bomb. The project was, with respect to its own objectives, a success.

[57]   The proportions of the various isotopes of a given element in a material.

[58]   Who wouldn't date Dwayne Johnson, really?

two years of evaluating spectrometry data to have enough elements and come to a conclusion about the age of the earth. But there is another thing, and I'll never forgive myself if I miss this opportunity to pay him enough homage, because he is another planetary *rock* star.[59] While he was studying the isotopic composition of lead on the earth, Pat noticed that the levels of lead on the surface of the earth and in the atmosphere had increased significantly as industry expanded. He spent the rest of his life warning the public about the dangers of lead, particularly in a 1965 article titled "Contaminated and Natural Lead Environments of Man,"[60] in which he wrote that lead had entered the food chain. He wanted to have lead banned from industrial use, especially in the gasoline industry. He went up against the lobby for a company, Ethyl Corporation, that specialized in gasoline additives—just to reassure you conspiracy theorists, it was simply a regular lobby, which means a group that advocates for a cause . . . in this case, savings from gasoline additives. The lobby engaged experts of their own who stated there was no proof to demonstrate that leaded gasoline caused a danger of any kind. Patterson studied skeletons that were more than fifteen hundred years old and compared the level of lead in them with the level of lead in recent skeletons. He concluded that the concentration of lead had increased by one thousand times over the course of fifteen hundred years. He was absolutely certain of the danger—especially because we've known about lead's toxicity since antiquity.

After many years of struggle—in 1971, he was unjustly excluded from a research panel on lead pollution in the atmosphere, even though he was one of the world's greatest experts on the subject—he ultimately won the battle in 1973 when the US Environmental Protection Agency recommended a 60 percent reduction of lead in gasoline and completely phased it out by 1986. In 1978, Pat Patterson was finally asked to be on the National Research Council, and he was instrumental in having lead eliminated from paints, varnishes, packaging, and water distribution systems. Today, almost all of his recommendations have been implemented.

---

[59] Pun definitely intended.

[60] *Archives of Environmental Health* 11 (1965): 344–60.

32. Mars

The Red Planet Mars—actually, it's more of an orangish color—has fascinated humans ever since we figured out how to get a better look at it. We think it has so many similarities to Earth—kind of rude since we now know that Venus, beneath its thick layer of clouds, is far more similar to Earth, but never mind. The fascinating thing about Mars, or at least what makes this planet so popular, is not its geology, its atmosphere, or its ice caps: It's the Martians. Because, now, prepare yourselves . . . you are now going to learn the truth about what governments have been hiding from you for more than 150 years: There was a civilization that . . . No, I'm joking. That was a cheap joke, I know. I'm sorry.

After a few centuries of observing Mars by telescope—the first observation was in 1610, by Galileo—astronomers had gained a fairly accurate idea of the planet's areas of albedo features.

Albedo Features

Albedo is the reflectiveness of a surface. It ranges from an albedo of 0, for a surface that doesn't reflect any light, to an albedo of 1, for a perfect mirror that reflects absolutely all light radiation. Of course 0 and 1 are ideal values; you won't come across 0 or 1 at every street corner. Basically, albedo indicates the shiny property of an object. Thus albedo features are bright areas that contrast against surrounding areas. When studying the surface of a body in space—*exogeology,* aka planetary geology—these features are the easiest ones to detect. The lunar "seas," darker areas on the moon's surface, are excellent examples of albedo features.

By the nineteenth century, astronomers had a pretty good idea of what was shiny or not on Mars, and they began to dream of seas of liquid water on the planet. When an error led them to believe they had detected the spectroscopic signature of water in the atmosphere around Mars, interest in the possibility of life on Mars grew and spread to the general public. The American astronomer Percival Lowell, who discovered Pluto, the planet that's not a planet—we'll get to that— thought he could prove the existence of a Martian civilization based on the observation of artificial canals on the Red Planet. Actually, these canals were first observed by Giovanni Schiaparelli in 1877 and were later found to be optical illusions.

Part of being human is seeing what we want to see, and as it happened, on July 25, 1976, the *Viking 1* space probe[61] flew over Cydonia Mensae, a rocky massif (mountain mass) in the northern hemisphere of Mars and took a photo. This photo first stunned the scientific community, then the whole world! A face. Undeniably, a face. Carved in rock and at least 1.2 miles long and a quarter of a mile high. Such a structure couldn't possibly be natural! But a few years later, better photos of the area definitely showed that the face was only a trick of the light and shadows on the rocky outcrop. When clearly illuminated, it looked nothing like a face. So why did we see a face? Is it really human nature to see what we want to see? In a way, yes, because of the well-known, but not very well understood, phenomenon called *pareidolia*.

Pareidolia

Pareidolia is an optical illusion that tricks the brain more than it tricks the eye. It is interpreting a blurred or ambiguous image as a clear and perfectly defined shape—usually a face, a person, or other living creature. Seeing a rabbit in a cloud is pareidolia. The human brain, through many generations and over the course of hundreds and thousands of years, evolved the ability to identify the slightest danger and created multiple defense mechanisms. One is the ability to rapidly spot things that represent a potential threat in a given environment, to tell the difference between what is an animal and something that is just moving. Similarly, the ability to quickly identify a face might help in seeing that it has an angry expression, and therefore is potentially a threat.

The entire process isn't completely understood. But when our brain is presented with an unfamiliar perception, our first reaction is to try to equate it to a familiar perception—when it comes to shapes, the temporal lobe is involved. This is the mechanism that lets us see a colon and a parenthesis as a smiling face. :)

It's also the mechanism that makes a mountain on another planet bathed in partial light look like a face.

---

[61] *Viking 1* orbited Mars for more than a month.

So, no life on Mars? It's highly unlikely. Currently, Mars is the second-most understood planet in the solar system, behind only Earth. Hundreds, thousands of square miles of its surface have been studied by orbiting probes and by rovers rolling around on its soil, yet nothing indicates the possibility of life existing on Mars. Nothing . . . well, almost nothing.

In 1965, *Mariner 4* was the first space probe to orbit our red neighbor in space. Its first observations seemed to leave no doubt: No ocean, no rivers, no liquid—the planet was completely arid. Furthermore, it had no magnetosphere—it has no magnetic field—which makes it defenseless against cosmic radiation and solar winds. Last, its atmosphere is not very dense; the atmospheric pressure is nearly 170 times less than that of Earth, making it impossible to have liquid water. Therefore, our understanding of life forced us to think that if there were life on Mars, it could only be bacterial life, or more generally, in the form of single-cell organisms. The goal of the *Viking* program in the 1970s was to find these organisms. And among all the experiments installed on the *Viking* probes, only one seemed to provide a positive finding, the *labeled release experiment*.

Now we've covered Mercury, Venus, our good ol' Earth, and Mars, but before we get to the gas giants, we find a large area where for a long time we thought we had a planet, then four planets. Then we realized that, for one thing, they weren't planets, and for another thing, there were more than four. Way more!

Labeled Release

The labeled release experiment consisted of taking a soil sample and placing it in a favorable environment for bacterial growth—rich in water and nutrients—then monitoring it for possible reproduction. When the experiment was performed, they found an increase in the production of carbon dioxide, which can be interpreted as a result of respiration, even though no organic molecules were found.

Today, the general scientific consensus is that the production of carbon dioxide can be explained by nonbiological processes, but there are still many different interpretations of the results.

33. **The Missing Planet**

In 1766, the German astronomer Johann Daniel Titius discovered a mathematical curiosity about then-planets in the solar system. He noticed that the average distance from the sun seemed to follow a relatively simple mathematical rule. We talk about the *average* distance from the sun, because the orbits are usually an ellipse; sometimes planets are closer to the sun and other times farther. Using the average distance accounts for that, and nobody gets annoyed. Counting by astronomical units—the average distance between the sun and Earth—the nth planet's average distance from the sun is 0.4 au + 0.3 × $2n$-1. By setting Mercury at a distance of 0.4 au from the sun, for the rest of the planets, we find:

Venus (n = 1) 0.4 + 0.3 = 0.7 au,

Earth (n = 2) 0.4 + (0.3 × 2) = 1 au,

Mars (n = 3) 0.4 + (0.3 × 2 × 2) = 1.6 au,

Jupiter (n = 5) 0.4 + (0.3 × 2 × 2 × 2 × 2) = 5.2 au,

Saturn (n = 6) 0.4 + (0.3 × 2 × 2 × 2 × 2 × 2) =10 au,

etc.[62]

By the way, before Titius said anything, the Polish philosopher Christian Wolff had already noticed this series of numbers, but he didn't write an equation for it. He, by the way, had merely cited the Scottish mathematician David Gregory, who was apparently the first person to discuss the subject in 1702. Yet this law is attributed to Johann Elert Bode, also a German astronomer, because in 1772 Bode included it in his *Instruction for the Knowledge of the Starry Heavens*.[63] The law is more commonly known as the Titius-Bode law. And this law posed a problem—oh goody, another problem!

[62] The following can be expressed more simply if you count by ten astronomical units; the successive numbers would then be: 0, 3, 6, 12, 24, 48, 96 . . . to which you add 4: 4, 7, 10, 16, 28, 52, 100 . . .

[63] *Anleitung zur Kenntniß des gestirnten Himmels.*

The very observant among you may have noticed that in the number series above, we skipped right from the third spot (Mars) to the fifth spot (Jupiter). That's because there was no known planet located between Mars and Jupiter. But in 1781, William Herschel, a German-British astronomer and composer—yes, composer—discovered a new planet, Uranus. And guess what? Uranus was sitting at an average distance of 19.2 au, right where the Titius-Bode law predicted a seventh planet! "Bam!" So, Bode got up some nerve and suggested there was an unknown planet between Jupiter and Mars just begging to be discovered. He used his formula to calculate its average distance from the sun and found that the missing planet should be located approximately 2.8 au from the sun.

In 1800, Franz Xaver von Zach invited some astronomers to meet in Lilienthal, Germany. Those in attendance agreed to look for this unknown planet. They knew what distance to look at; the "only" thing left to do was search the sky. They didn't find it. But in 1801, Giuseppe Piazzi, a Sicilian astronomer and director of the Astronomical Observatory of Palermo, looked at the sky to observe a particular star, the eighty-fourth star listed in Nicolas-Louis de Lacaille's catalog of zodiac stars. Instead, Giuseppe discovered an object that didn't follow the same path as the other stars. He "quickly" ruled out the idea that it was a comet—by taking twenty-four observations between January 1 and February 11. He announced he had discovered something sort of like a comet, but he added:

> Because of its slow uniform motion, it occurred to me, after repeated measurements, that it could be something better than a comet.[64]

Piazzi named it Ceres, for the goddess of agriculture from ancient Roman mythology—Demeter or Demetra to the Greeks—who just coincidentally happened to be the patron goddess of Sicily.

64   E. G. Forbes, "Gauss and the Discovery of Ceres," *Journal of the History of Astronomy* 2 (1971): 195.

Before other astronomers could confirm the existence of Ceres, its orbit brought it too close to the sun, making it impossible to observe. They had to wait several months for confirmation. And there was an additional challenge; after several long months, it would be hard to tell where to find this celestial body. Remember, Ceres had been discovered by accident, because it is so small. Theoretically, the problem could be handled mathematically using the few existing observations to determine the orbit. Our dear friend Carl Friedrich Gauss—whom we met earlier in the discussion of electromagnetism (see page 86)—jumped on the problem and developed a method to describe a body's orbit using only three observations. Gauss made his predictions and sent them to Zach and to Heinrich Olbers. So they found themselves searching in the sky for a tiny body—one they weren't sure really existed—in a location they were unsure about, because they didn't know if the calculations were correct—but, well, still . . . it was *Gauss*. The astronomers did succeed in finding Ceres in the expected location on December 31, 1801. This confirmed both Gauss's mathematical method *and* the existence of Ceres. Confirming the discovery of Ceres took an entire year—and also confirms that astronomers can think of nothing better to do on New Year's Eve.

Ceres was considered a planet, the famous "missing planet," and was duly included in astronomy textbooks. It was taught that the solar system included (starting closest to the sun) Mercury, Venus, Earth, Mars, Ceres, Jupiter, and Uranus.

34. Pallas, Juno, Vesta, and Everyone Else and Their Mother

A few months after the existence of Ceres was confirmed, on March 28, 1802, Olbers attempted another observation, and just by chance another body passed in front of Ceres—peek-a-boo. This body *also* moved in the celestial vault. And that's when the astronomers became—let's say—perplexed. They understood the logic that allowed them to envision the existence of a planet between Mars and Jupiter, but two planets . . . that was curious. You might say Pallas—that's its name—had already been discovered in 1779 by Charles Messier, a French astronomer and comet hunter, but he made only one observation; he thought it was a star—he quickly realized it wasn't a comet, thought "so who cares," jotted it down, then went on to other things. OK, so, getting back to the solar system: Mercury, Venus, Earth, Mars, Pallas, Ceres, Jupiter, Saturn, and Uranus.

On September 1, 1804, Karl Ludwig Harding, a German astronomer, discovered in that area a new planet, which he named Juno. And then a trend started all over the world. OK, good, the solar system: Mercury, Venus, Earth, Mars, Juno, Pallas, Ceres, Jupiter, Saturn, and Uranus. Then in 1807, on March 29, Olbers discovered a fourth planet, Vesta, not too far from there. Let's go over the solar system as it was taught then: Mercury, Venus, Earth, Mars, Vesta, Juno, Pallas, Ceres, Jupiter, Saturn, and Uranus. Eleven planets. The four most recently discovered planets were small and light, but they were certainly planets. No problem. Up until Astraea.

In 1845, Karl Ludwig Hencke—yes, Karl Ludwig was obviously a very popular name in Germany; apparently their equivalent of Emma or James at the time—was an amateur astronomer who worked at the post office. He sat down to look for Vesta, and, like the professional astronomers before him, he discovered a new body: Astraea.

And in 1847, the list grew: Hebe, Iris, Flora; with Metis in 1848; and Hygeia in 1849. This just wasn't working for the scientific community anymore; they had to make a decision. There were obviously a lot more bodies between Mars and Jupiter than anyone had previously thought. At the same time scientists needed to understand what these objects were, where they came from, and how to classify them.

For one thing, the four first "planets" discovered between Mars and Jupiter weren't in the right places with respect to where the orbit of the missing planet should have been. And on top of that, they were starting to find rocks—sorry, but there's no other word for them, at least not yet—by the dozens. As early as 1802, William Herschel suggested the term *asteroid*—literally "star shaped"—because they looked like stars. In 1850, the notion of an asteroid belt came along. And over the next few decades, more and more asteroids were discovered within a 110 million-mile-wide band between Mars and Jupiter. Vesta, Juno, Pallas, and finally Ceres—the largest known asteroid—were demoted, one by one, from their planet status. The issue came up again in 2006, but you'll hear more about that when we talk about Pluto.

Today, discovering a new asteroid is a nonevent; between 1995 and 2005, dozens of asteroids were discovered daily. As of 2015, the number of *identified* asteroids was over 580,000 —there are also a number of unidentifiables; the tiniest ones are the size of a grain

Hellacious Obstacle Course . . . or Not

Contrary to what we see in movies, crossing the asteroid belt doesn't require the skill of a Jedi knight. In fact, in the asteroid belt, the asteroids are hundreds of thousands to even millions of miles apart. At our scale, they are actually rather sparse.

of sand—and there must be millions more asteroids waiting to be identified.

While some things are still uncertain about the formation of the asteroid belt, there are generally two schools of thought: the Scientists and the Lampposts—I'll use the term *Lampposts*, because I want to be sure I'm not insulting nor mocking its supporters. Scientists think the asteroid belt was formed at the same time as the solar system, as a large band of dust and rocks spread throughout an orbit—exactly like it was for the terrestrial planets—except for Jupiter's massive presence; all that gravity prevented the formation of the potential planet by causing collisions and by snatching material away from the developing planet. It's quite likely that Ceres, with its diameter of thousands of miles, was a protoplanet in our system. By accretion with other rocks, Ceres could have become a planet in its own right—and would have filled the position of the famous missing planet.

As for the Lampposts, they think that such a planet did once exist, that it was destroyed by collisions and by loss of material caused by Jupiter, and that the asteroid belt is the debris from this planet. However, this isn't the idea that earns them the nickname Lampposts. Today, the total mass of the asteroids in the belt is much too small to equal the mass of a properly formed planet, but one could argue that the majority of the asteroids were expelled from the area, whether by solar winds or by the gravity of Jupiter. Nope, that's not the issue.

Here is what drives me to discuss these Lamppost folks: It's because it is "necessary" for them to think a number of things that make absolutely no sense. First, life existed on the planet, and this life-form was intelligent. Sufficiently intelligent to master atomic energy and destroy their own planet with atomic bombs. Then it seems, according to the Lampposts, this propelled another planet, Nibiru, farther into space. Nibiru is a yet undiscovered planet presumed to be orbiting beyond Pluto at present—the famous planet X that we'll talk about when we discuss Pluto. *And* Nibiru passes close to Earth every thirty-six hundred years. That's when life on the surface of Nibiru had the chance to and did indeed move to Earth to become our ancestors, or inhabitants of Atlantis and the lost continent Mu.[65] These theories are widespread on the internet and are presented in the Comic Sans MS font, which, as everyone knows, is certainly the most serious and intellectual font in the history of typography.[66]

Beyond the asteroid belt, we find the large gaseous planets, aka the gas giants, starting with the largest one of all: Jupiter.

35. Jupiter

With a diameter that's eleven times Earth's diameter, Jupiter is the fattest planet in the solar system; it is the largest and most massive of any of the planets. In fact, if you put all the other planets in the solar system together, the mass of Jupiter is approximately two and a half times *more* than that. Jupiter is so massive that the center of gravity between Jupiter

---

[65]　It is fascinating to note that for the Lampposts, the fact that no one has found Nibiru is proof that it exists.

[66]　There are numerous sites, petitions, and various policies aimed at eliminating this font. It is widely considered the ugliest and most ridiculous of all character fonts.

and the sun (meaning the barycenter it orbits around) is actually outside the sun—barely, but still . . . In other words Jupiter is so massive it makes the sun orbit around that center of gravity. Jupiter moves the sun. The sun's diameter is about ten times larger and its mass is one thousand times greater—as the sun goes, so goes the solar system.

Anyone who looks at a photo of Jupiter and then tries to draw it draws wide reddish stripes and a huge rust spot that looks like some cosmic eye. That's the famous Great Red Spot of Jupiter. It's the largest and longest-running cyclone in the solar system—technically, it's an anticyclone. It's easily larger than the whole planet Earth, and its winds blow at 430 mph! The Great Red Spot was discovered by Giovanni Domenico Cassini, an Italian astronomer. When he became a French citizen, he took on the new name of Jean-Dominique Cassini.

Jupiter has been known "since time immemorial," as they say. The Babylonians knew about Jupiter, as did the Chinese, and the Egyptians, and the Greeks. That's because Jupiter is visible to the naked eye. It's one of the seven astronomical bodies visible to the naked eye that don't turn with the celestial vault—one of those wandering stars: the sun, the moon, Mercury, Venus, Jupiter, and Saturn. It wasn't until 1610 that Jupiter was first seen through a telescope by Galileo himself. He saw Jupiter's four moons—the Galilean moons. This discovery—the first-ever sighting of a celestial body revolving around a celestial body that wasn't the earth—convinced him that the earth wasn't the center of the universe and that the earth revolved around the sun.

Even though Jupiter can't beat Saturn in the ring category, it too has rings. They were observed for the first time by the *Voyager* space probe in 1979. In 2007, the *New Horizons* probe photographed the rings seven years before the Pluto flyby.

Jupiter's composition is similar to that of the sun and the stars: a lot of hydrogen (75 percent), less helium (24 percent), and even less of the heavier elements (1 percent). But Jupiter isn't massive enough for its own gravity to trigger the thermonuclear fusion process. Jupiter remains a star wannabe. Be that as it may, recent studies suggest the frequent, violent, high-altitude storms break down the methane molecules and make it rain carbon atoms. As they fall, they are subjected to more and more pressure, because the atmospheric pressure increases as you get closer to the surface. The carbon becomes graphite and eventually is compressed into solid diamonds. Literally, raining diamonds! Alas, the treasure that's sleeting is also quite fleeting. At an even lower altitude, the pressure makes the diamonds heat up above their melting point and they become liquid. Sadly, the diamond rain goes down the drain. By the way, Saturn also experiences this priceless rain.

On top of all that, here's the most interesting thing about Jupiter: The planet and its moons constitute an entire system, the Jovian system.

36. The Jovian System

It all started when Galileo discovered four moons around Jupiter—we call them the Galilean moons. They are Io, Ganymede, Callisto, and Europa. The closest moon to Jupiter is Io; it completes its orbit in only forty hours. Io is approximately the size of our moon—5 percent larger, but 20 percent heavier. Because of the speed at which Io travels around Jupiter and because it's so close to Jupiter and its gravitational field, Io experiences tidal forces that cause the most extreme volcanic activity in the solar system—Io is the second-most active object in the solar system, second only to the sun.

Tidal Forces

The moon's gravitational force on Earth attracts not only the land but the oceans as well. Since oceans are fluid, the moon's pull causes a fluctuation in sea level. The rhythm of the tides depends on the rotation of the earth as well as the moon's revolving around Earth. In general, all massive bodies exert gravitational attraction that can cause tides in oceans and in molten magma, for that matter. These tidal forces can lead to some pretty violent volcanic activity. And that's what's going on with Jupiter's moons—Io in particular.

On Io, more than four hundred active volcanoes spew sulfur compounds nearly three hundred miles up in the air—I mean, if there were any air; Io has almost nil atmosphere. The sulfur lying all over the ground explains Io's orange-yellow color. The constant volcanic activity explains why the moon's surface looks so young: because it is constantly being covered with new material. Or maybe Io is just simply young; that's also a reasonable hypothesis.

That same day, January 7, 1610, Galileo discovered the second of Jupiter's moons, Ganymede. It's the only moon in the solar system that has a magnetosphere—it generates its own magnetic field. It's also entangled within Jupiter's huge magnetosphere. Ganymede has a magnetosphere, because it's fully differentiated. Its liquid core is rich in iron and in constant motion due to convection—the core's movement generates the magnetic field.

Callisto was also discovered in January 1610—also by Galileo, though it might have been identified by the Chinese astronomer Gan De in the year 362; if that's true, it was quite an accomplishment. Of the four Galilean moons, Callisto is the farthest away from Jupiter. So it's less subject to Jovian tidal forces, and as a consequence, it may be only partially differentiated. The *Galileo* probe indicated that Callisto most likely has a silicate core and an ocean of liquid water sixty miles below its surface of ice and rocks. And when I say "liquid water," understand this: If that ocean

Planetary Differentiation

During the formation of a planet—or any celestial body—when the inside of the body melts, the elements that are more dense sink toward the center. At the same time, elements that are less dense "float" toward the outer layer. When the planet cools, this structure will be preserved, with a molten core at the center, surrounded by a mantle, surrounded by a crust. When the structure that results is in distinct layers, we talk about a differentiated planet. For rocky bodies, you usually need a relatively high temperature—on the order of 1,800°F—to get a gooey center. On the other hand, icy bodies need only to reach the melting point of water to have a liquid core.

exists, it contains more water than we have on the whole planet Earth, even though Callisto is only about one-third the diameter of Earth. Plus when you hear "liquid water," it means perhaps, in this ocean, life as we know it might exist. Except at sixty miles below its surface. So, it doesn't make a very good swimming hole. Callisto remains, to this day, the most suitable of all Jovian moons to serve as a human base for exploration of the Jovian system. Unlike Io, Callisto's surface has craters scattered all over it. This tells us two things: (1) There's no tectonic activity, and (2) it must be very old—it could date back to the formation of the solar system, 4.5 billion years ago.

And finally, Europa—I saved the best for last—a very special moon for several reasons:

Its icy surface makes it the smoothest moon in the solar system.

Liquid water geysers have been detected on its surface.

Europa may be hiding a planetary ocean that is fifty-five miles deep.

If you want to bet on the planet most likely to have life as we know it, put your money on Europa. NASA plans to send a probe there in the next few years to learn as much as possible about this shiny moon.

Space Odyssey

Arthur C. Clarke, the author of multiple best-selling science fiction books, wrote one of the greatest epics of sci-fi literature, a tetralogy whose first volume was adapted—well, not so much an adaptation, as a book written concurrently with the development of the film script—by Stanley Kubrick for a film that is considered one of the best movies of all time—*2001: A Space Odyssey*. In the later books, Europa plays a big part and is now eternally associated with the warning at the end of the second book, *2010: Odyssey Two*:

ALL THESE WORLDS ARE YOURS—EXCEPT EUROPA. ATTEMPT NO LANDINGS THERE.

In addition to the four Galilean moons, Jupiter currently[67] has sixty-five more moons, bringing the number of moons in Jupiter's Jovian system to a grand total of sixty-nine (fifty-three confirmed and sixteen unconfirmed or "provisional") moons and an awesome set of rings. Like I said before, it's a whole system unto itself.

37. Saturn

Immediately recognizable by its rings, Saturn is the second fattest planet in the solar system. Saturn is ninety-five times heavier than Earth, but occupies nine hundred times the volume. One of the planets that's visible to the naked eye, it's been observed since prehistoric times. In 1610, Galileo was so excited about the improvements he'd made to his astronomical telescope that he spent his time observing everything he could and taking notes; when he looked at Saturn, he couldn't see it clearly enough to tell what the rings were. He described the planet as having "ears." Needless to say, Galileo had no idea what he was looking at and sounded for once—forgive me for being blunt—like a knucklehead. As it happens, at the time Galileo saw this, Earth was passing even with the edge of the rings; he was looking at them from the side. And while the rings are indeed very wide, they are also rather thin—some as thin as 32 feet—for rings that are visible with something as simple as amateur telescope or a good pair of binoculars. Then because he was viewing the rings' edge, the rings completely disappeared for a while and then reappeared several months later—talk about confusing!

Because its rotation is so fast, Saturn is 10 percent wider than it is tall—as if *wide* and *tall* mean anything in space; let's just use the poles as the vertical axis of the planet. The wind on Saturn's surface travels at 1,100 mph, making Jupiter seem like a tranquil wasteland by comparison. At the north pole, two phenomena occur: First a strange permanent cyclone called the polar vortex, awesome, like none seen elsewhere in the solar system. Then, there's this abstract hexagonal structure on the north pole—I use *abstract* in the sense that it's not supported by any

[67]   The latest count as of 2017, but there's nothing stopping us from finding more.

actual physical structure or terrain—and as if that were not bizarre enough, even though masses of gases are moving in circles around the pole, the "structure" doesn't turn. It may actually be an illusion caused by standing waves, where displacement propagates in a way that offsets the winds, giving the waves the appearance of standing still.

One last thing about Saturn . . . it's not a very important point, but it's often misinterpreted: The average density of Saturn is less than the density of liquid water; thus some conclude, Saturn would float in water. Two things about this: First, it's the *average* density of Saturn, and even though its density at the surface is definitely less than water's density, its density below the surface is much greater. Secondly, if we tried to amuse ourselves by putting Saturn into one of Earth's oceans, our entire planet would be crushed by Saturn's immense gravity. Remember, Saturn is ninety-five times more massive than Earth. It's more important to understand that *if* you have a liquid ocean large enough, and *if* you somehow manage to put Saturn on the surface of this ocean, and *if* the ocean isn't immediately sucked up by Saturn's gravity, *then* Saturn would float. It's a thought experiment. That's all.

That's the last item about Saturn, but we still need to talk about its moons. By the way, there's no final decision about the sixty-nine moons observed; to date we've only confirmed the existence of fifty-three of them. Most of them are very small, only about thirty miles across. Although seven of them are really exceptional—Titan, Mimas, Enceladus, Tethys, Dione, Hyperion, and Phoebe—there are three that I especially want to talk about: two because they are interesting from a scientific point of view, and the third for a very silly reason; but I do what I want, because it's my book.[68]

38. Mimas, Enceladus, and Titan

Let's start with the silly subject first, Mimas. Discovered in 1789 by William Hershel, this moon is nearly 250 miles in diameter and is primarily composed of frozen water with a little bit of rock. Mimas is notable because of

[68]   So there!

an immense crater—immense compared to the moon's size—that lets us see classic crater characteristics: It's round, it has an elevated rim nearly three miles high, and in the center you find a peak, slightly higher than its rim—almost six miles high. It clearly resembles what it looks like when a drop of water hits the surface of perfectly calm water. But what really gets me about Mimas is its striking resemblance to the Death Star in *Star Wars*.[69] There it is. That's all I wanted to tell you about Mimas.

Enceladus is seriously *beaucoup* more interesting. It ties with Titan as Saturn's most intriguing moon. Enceladus is of modest size, with a diameter of three hundred miles. Enceladus has some internal activity that, even today, no one understands. Still, Enceladus is to Saturn what Europa is to Jupiter—perhaps even more so. It's been determined that Enceladus has a liquid ocean under its icy crust, and it has geysers. The geysers indicate that there's definitely activity—and it's definitely hot—and from what we see expelled from the geysers, there are organic molecules. Maybe even more than Europa, Enceladus might make us hope: heat + liquid water + organic molecules = the ingredients for life as we know it. On top of its icy surface, Enceladus seems to have a layer of water-based snow, about 330 feet thick. That means it's been snowing on Enceladus for at least one hundred million years, and therefore, the heat source that produces the geysers has been active for at least as long. After the sun, Enceladus is the brightest object in the solar system, because its surface is covered with "clean" ice—more precisely, it's the object with the highest albedo. Alas, Enceladus is too small to be seen as bright from Earth. Finally, Enceladus does have an atmosphere—a thin one, primarily found only in the polar areas, but an atmosphere nonetheless—made mostly of water vapor.

Titan, on the other hand, as far as atmosphere goes, is a different story. The first moon discovered orbiting Saturn—by Huygens in 1655—Titan is the second largest moon in the solar system after Ganymede. It's a little larger than Mercury, and it's half water and half rocks. The unique thing about Titan is that it's the only moon in the solar system that has a truly dense atmosphere. The thickness of the atmosphere ranges from

[69]   As if I had to tell you that the Death Star was in *Star Wars*.

125 to 500 miles high—where Earth's is usually around 60 miles high. Titan's atmosphere is too opaque for us to get a good look at its soil to study it properly. The atmosphere's density at the ground is 1.5 times Earth's and is composed of 98 percent nitrogen, plus other stuff. Titan is the only body in the solar system, besides Earth, with an atmosphere that is primarily nitrogen. Earth's atmosphere is 78 percent nitrogen, plus oxygen and other stuff. In 2005, the *Cassini* probe dropped the *Huygens* landing module on Titan to take photos of its surroundings when it reached the ground. We've come to find out that if you replace Earth's water with methane, the environment on Titan is not all that different from Earth's. It may rain methane or methane may blow around on Titan's surface. The cryovolcanic activity—like volcanic activity on Earth, except these eruptions spew forth water, ammonia, and methane—shows a significant presence of water on Titan.

Needless to say, we still haven't discovered everything in the solar system, on the planets and all those moons, so we keep searching everywhere for life—especially life as we know it.

But, wait! There's more!

39. Uranus and Neptune

Uranus was the first planet discovered that wasn't visible to the naked eye, and it quite suddenly expanded the size of our solar system. At the time of its discovery, we had already discovered moons around Jupiter and Saturn, and we were about to discover the asteroids we thought were planets, but Uranus was farther out. Uranus changed the way people—the species, not the magazine—thought about the entire universe, and especially the solar system. If I've put Uranus and Neptune in the same chapter, it's because their stories are indissociable.[70]

Uranus was discovered by William Herschel in 1781 (you remember Herschel—I mentioned him on page 128). Even though Uranus had been seen on numerous occasions before the eighteenth century, the astronomers had mistaken it for a star. Uranus is so dim and so slow—no

---

[70] Wow! Put that on your SAT vocab list for "cannot be separated."

offense intended—that it was very difficult to identify Uranus as a planet. For example, John Flamsteed, an English astronomer, observed Uranus several times in 1690 and cataloged it as a star named 43 Tauri. Pierre Charles Le Monnier, a French astronomer, observed it regularly between 1750 and 1769, too short of a period to notice the heavenly body move in the starry sky. But in the end, it was Herschel who found and identified it as a planet, even though he wasn't looking for it at all. This was starting to be a habit: Apparently serendipity[71] rules in astronomy.

Herschel's thing was double stars. He looked for them everywhere.

He was trying to measure the parallax (see page 107) of stars by changing his telescope's eyepiece, which changes the magnification. During one of his observations, he noted a small dot that seemed to be emerging from behind Saturn. So he changed the eyepiece of his telescope several times. The distant stars didn't look any different; their size didn't change. In contrast, the dot grew larger with each magnification, indicating close proximity. That told him it wasn't a distant star. He noted his observation in his journal: It was a curious thing . . . a nebula, or maybe a comet. He alerted the scientific community about his discovery, most likely a new comet.

Double Stars

A double star, or *binary star system*, consists of two stars— no kidding!—that are close enough for their gravity to affect each other. It's a very common arrangement in the universe— there's ongoing debate about just how common. There are also triple, quadruple, quintuple, even sextuple—like Castor.

Though Herschel was particularly conservative regarding the nature of the object, some astronomers started thinking planet. Such was the case with Bode, who felt sure it was a planet. After all, the object was right where his famous law (see page 127) said one should be. The Russian astronomer Anders Lexell also thought it was a planet, because he thought a comet would have a closer perihelion[72] than he calculated

---

[71] Pleasant, unexpected discovery.

[72] Come on now, don't be afraid of *perihelion*. We talked about perihelion on pages 103 and 104.

for this object. In trying to solve the mystery, astronomers calculated the path of this "comet," which they could do very well, but it didn't come out right. The calculations didn't match the observations. Charles Messier—remember him, the comet specialist?—had a look at it with his telescope, and he found that the object looked more like a circle, like Jupiter, than like a comet, and Messier knows comets. So, Lexell calculated the trajectory of this object as if it were a planet, and miracle of miracles! The calculations and the observations matched really, really well!! It didn't take long for the scientific community to officially recognize the object as a new planet. Later, a lot later, on March 10, 1977—by the way your humble servant was nine years old at the time—they discovered Uranus also had rings!

Now for a little detour about the name of the planet—some folks will give me a hard time if I don't tell you about it. After its discovery in 1781, King George III, the king of England, offered Herschel an annual salary on the condition that Herschel come and live at Windsor, so the royal family could observe the skies using his telescope, and Herschel gladly accepted. So when he was asked to choose a name for the planet, he wanted to name it Georgium Sidus,[73] or Georgian Planet. Surprisingly enough, the name wasn't all that exciting to folks outside of Great Britain.[74] The Swedish astronomer Erik Prosperin proposed Neptune, and Bode suggested the name Uranus, for the ancient Greek god Ouranos, god of the sky. To Bode, the name seemed appropriate because Saturn was the father of Jupiter, and Ouranos was the father of Saturn. Ultimately, seventy years after its discovery, in 1850, the name Uranus became the official name of the planet—and with the accidental slip of the tongue it becomes *your anus*, and that too is part of the legend.[75]

---

[73]   George's star.

[74]   For one thing, 1781 was right in the middle of America's Revolutionary War with England and King George III.

[75]   Think about it! A whole new galaxy of jokes was born: "Jupiter, Saturn, your anus" or "Your anus is 1.7 billion miles from Earth." For many folks, it's a very effective mnemonic.

Studying your anus posed a problem, because you got a stiff neck. . . . No wait. . . . The study of Uranus posed a problem—as so often happens in science, when you say *problem*, you mean "future discovery"—because its orbit is pretty strange. First of all, its axis of rotation is completely whacked; its poles—defining the axis of rotation—are just about lying in its own orbital plane. To get an idea of what I'm talking about, most of the planets are like battling tops spinning on a table—let's call the table the ecliptic plane—and Uranus doesn't spin like a top; Uranus is on its side rolling like a wheel!

Galileo had noticed Neptune while observing Jupiter in 1612, but he thought it was a star and left it at that. In 1796, Neptune was also observed by Joseph Jérôme Lefrançois de Lalande and by John Herschel—the son of William, because, yes, sometimes, it's in their genes—but he didn't draw any conclusion about it. It took an international team effort from people with a variety of skills to discover Neptune. And the discovery was made in trying to solve problems with Uranus's orbit. As early as 1788, astronomers had problems with its orbit, and as time passed, they saw increasing differences between the expected behavior and actual observations. Several astronomers tried to explain the phenomenon, theorizing about the gravitational influence of Saturn and Jupiter, but it was a puzzle they just couldn't solve. The French astronomer Alexis Bouvard offered a table of expected locations for Uranus, but that was no good, either. Then it occurred to Bouvard, the problem might be the influence of an unknown planet beyond Uranus.

Independently, the French mathematician Urbain Le Verrier and John Couch Adams, an English mathematician, went to work on calculating the characteristics of the possible planet and to determine its probable location. Though there were plenty of errors in their calculations, they each suggested the same location, also independently. Adams tried to get an English astronomer to look in his suggested direction, but there was a delay.

Johann Gottfried Galle, a Prussian astronomer and friend of Le Verrier, took the results of Le Verrier's calculations and pointed his telescope in the suggested direction. After looking for several hours, he eventually found Neptune where Le Verrier said it would be—not exactly but close enough, apparently. If you think about it, the eighth planet of the solar system was found on a piece of paper by a mathematician: That was a first! But as astronomers began to study Neptune, they realized this planet, so similar to Uranus, also had a whacko orbit—discussed as Neptunian perturbations. Less than with Uranus, but still.

Well, the astronomers went apeshit.[76] There was obviously a new planet to discover, and the Americans, a bit behind in the discovering planets category, flung themselves at the prospect, with Percival Lowell leading the pack.

40. Pluto, the Fallen Planet

Percival Lowell, who was previously interested in the artificial canals on Mars and who had devoted himself to proving the existence of Martians (see page 124), decided he would solve the puzzle of perturbations in Neptune's orbit. He assumed the cause of the perturbations was the same as that causing the perturbations in Uranus's orbit: another yet undiscovered planet, beyond Neptune, and he named it Planet X. Lowell searched for it for a long time and he never found it—mostly because it doesn't exist.[77] When he died in 1916, he left a significant portion of his wealth to the person who would continue the search.

Lowell's brother, Abbott, used the money to build a thirty-inch telescope for $10,000 to be operated by the American astronomer Clyde William Tombaugh. On February 15, 1930, Clyde took tons of photos. And in those images, he found a tiny dot that didn't move with the other stars. An American had finally discovered a planet!

---

[76] Technical terminology currently used to indicate genuine, unbridled, and extreme enthusiasm.

[77] Though for some folks—whom I nicknamed the Lampposts (see page 132)—the fact that we haven't found Planet X proves it does exist and that it is actually the planet Nibiru.

But this planet was too puny[78] to explain the perturbations in Neptune's orbit. Tombaugh concluded he had discovered a planet, but it wasn't Planet X. He called the planet Pluto—Disney changed the name of Mickey Mouse's dog from Rover to Pluto, in honor of the discovery. Tombaugh continued to search for Planet X, and he never found it, obviously—because it doesn't exist.

Even though Tombaugh never found Planet X, more than a thousand objects similar to Pluto were discovered during the twentieth century—they were similar in terms of size, mass, distance from the sun, etc. And astronomers started to get skeptical. It was a vague feeling at first—"Isn't this getting to be a bit much?"—then more familiar ("Haven't we been through this before with Ceres, Pallas, and that whole entourage?) then definite: "The situation is untenable; we can't adopt them all until ultimately the word *planet* has no significance." At that point, the astronomers were in Jenga mode, when the next move loses—and in schools, students learned the solar system as follows: Mercury, Venus, Earth, Mars, Jupiter, Saturn, Uranus, Neptune, and Pluto. Nine planets. First we have four terrestrial planets, then four gas planets, then . . . then another terrestrial planet, after the four giants of the solar system, and it is so small that its surface area can't even cover Asia. Pluto held on as long as it could in spite of a growing number of neighbors: Makemake, Orcus, Sedna, and others. The International Astronomical Union (IAU), the only organization officially responsible for classifying celestial bodies, didn't approach the question and didn't take a stance on the status of Pluto—an eminently political subject, made more so because Pluto was the only planet ever discovered by an American.

In 2003, however, an elephant entered the room; a new object had been discovered beyond Neptune: Eris. And Eris appeared to be larger than Pluto—not by much, but still. Since then, after the *New Horizons* probe passed through that area, we found out they are actually just about the same size—uh oh! Now what? If Pluto is a planet, then Eris should be a planet, too. IAU had to stop playing dead. They had to step

[78]   Such a perfect word for saying Pluto was too small and not massive enough.

in. No other option. So they took advantage of the opportunity to clearly define the status of a bunch of things in the solar system.

January 19, 2006, NASA launched *New Horizons* into space in the direction of Pluto. This space probe was the first probe to study the last planet in the solar system, the farthest and most "American" planet. *New Horizons* passed near Jupiter in 2007 and used Jupiter's gravity to literally fling itself toward Pluto, where it would arrive successfully in July 2015. Still, several months after the probe was launched, IAU announced a vote on a formal definition of a planet would take place at their next meeting in Prague in August 2006.

And on August 24, 2006, IAU proposed the following definition of a planet based on three points: a celestial body is a planet *if and only if:*[79]

> It has reached hydrostatic equilibrium (see page 95) without start-ing the thermonuclear fusion process.

> It is in orbit around the sun.

> It has cleared its neighborhood.

So in order to be a planet, a body's mass must be large enough for its own gravity to hold it in a compact ball. And it's essential that there is no thermonuclear fusion going on in its core—otherwise, it's a star. After that, the body must orbit around the sun. This part of the defini-tion presented a bit of an issue, because while we can certainly under-stand that it prevents us from considering the moon as a planet, it raises a question about the status of exoplanets[80]—some astronomers would have preferred for the definition to say the body had to orbit a star. Finally, the body must have "cleared up" its neighborhood, meaning that due to its gravity, it either collected the rocks, ices, and dust in its orbit or it threw them out of its orbit.

---

[79]   "If and only if" indicates it's a characterization: If all the conditions are met, it's
　　　a planet, and it's a planet only if all the conditions are met.
[80]   Planets of other solar systems.

And that's Pluto's problem. Because Pluto has nowhere near cleared its neighborhood. In reality, Pluto orbits in a zone that contains thousands of bodies. Pluto is definitely the largest known to date, but you'd be hard-pressed to say it had "cleared its neighborhood." So, on August 24, 2006, an IAU vote adopted this definition of a planet, and at that moment, Pluto fell from planet status. Pluto had officially become a *dwarf planet*, just like Ceres. American astronomers have tried a number of different things to get Pluto reclassified as a planet, but without success—they could just continue calling it a planet, because . . . well . . . you never know . . . it could happen.

While the definition adopted by IAU makes pretty good sense, one might question the way the vote was handled. It stinks of politics: In 2006, IAU had approximately 9,000 members; only 2,700 of them came to Prague for the festivities; the vote in question was held on the last day of the six-day event, and many members had gone home already; there were 1,000 people in the room at the time of the vote. Among those who were still there, some didn't approve of having such an important vote—the definition of what is a planet—with so few people. They abstained, hoping the quorum wouldn't be met and the vote would be invalid.

The number of votes was 424, and that was enough. Less than 5 percent of the voting members participated in the vote on the definition of a planet, which would downgrade Pluto from planet to dwarf planet. So, you can see why someone might question the validity of the vote.

The problem is, as I mentioned earlier, Pluto is not alone. And that brings us to what we find beyond Neptune, so we can wrap up our tour of the solar system.

41. Kuiper Belt and Oort Cloud

Today, it's agreed that the bodies located at a distance of between thirty and fifty-five astronomical units from the sun[81] are to be called the Edgeworth-Kuiper Belt, or more simply, the Kuiper (rhymes with *wiper*)

---

[81]   Or 2.8 to 5.1 billion miles from the sun.

Belt. This wide band in the solar system is similar to the asteroid belt, except it's twenty times wider and twenty to two hundred times more massive. There are at least three dwarf planets in the belt: Pluto, Makemake, and Haumea (Eris is located beyond the Kuiper Belt). But while the asteroids in the asteroid belt are primarily rocky, the bodies in the Kuiper Belt are made primarily of ice—astronomical ice, meaning water, ammonia, and methane. In this belt, we find more than seventy thousand objects that are over sixty miles wide. The Kuiper Belt is presumably one of the solar system's own reservoirs for comets.

Beyond the Kuiper Belt, you might find—I say "might" because it is still hypothetical—the Oort cloud. It's one thousand times farther away. If it exists, this sphere-shaped cloud is the largest reservoir of comets in the solar system. The Oort cloud is made of billions of bodies and extends from twenty thousand to one hundred thousand astronomical units from the sun.[82] This hypothetical bubble of icy bodies is the definitive boundary of the sun's gravitational sphere of influence. Farther out, we're no longer within the solar system, no matter how you look at it.

Comets

A comet is a small body of ice and dust orbiting around a star—in our case, the sun. Because a comet's orbit is very eccentric—mathematically, that means that the orbit is the shape of a long skinny ellipse (that is, the shape of a surfboard) sometimes the orbit brings the comet very close to the sun. The solar winds hit part of the body, giving it something kind of like an atmosphere (called a *coma*, or halo) and producing a long trail of ice and dust that flows with the direction of the solar winds (the *tail* of the comet). Because a comet is made mostly of ice, it has a high albedo. Comets shine so brightly that they can occasionally be seen with the naked eye—even during the day!

What makes comets leave the Kuiper Belt and the Oort cloud? Collisions, or maybe they come too close to a majorly massive body, and it knocks them out of their stable orbit. Then, they start to fall toward the center of the solar system.

---

[82]   Or 2 to 9 trillion miles from the sun, 1.5 light years being the outer limit.

Do you see how, all of a sudden, the scale blows up? We go past Pluto, the farthest "planet" in our solar system, and out to the Oort cloud, which is one thousand times farther away! It's hard to grasp the size of these distances without using a little analogy. I think it would be extremely useful to dedicate a small section to it. Before plunging into the vertigo of these astronomical distances, remember the entire solar system, as big as it seems to us, is ridiculously small at the scale of the universe. So small as to be completely insignificant, not like a drop of water in the ocean, but more like a drop of water in the solar system!

42. The Dimensions of the Solar System

It's always a little dicey to talk about the dimensions of the solar system, because it involves two extreme disproportions, that of distance and that of size. Making an analogy with the two side by side can quickly take us to scales that make no sense, like imagining an apple at the foot of the Empire State Building or a tiny green pea in the middle of Australia. So, I'll talk about distance and size separately.

OK, regarding size, we'll focus on the sun and its planets as well as the satellite we know best, the moon. The largest of these objects is the sun; let's represent that as 110 yards (100 meters) in diameter, so it's about the size of a soccer field or a football field. The second-biggest is Jupiter, with a 10-meter diameter, so it doesn't quite make it to the penalty mark, but it does make it past the 10-yard line. Next is Saturn at 27 feet (8.3 meters) in diameter. Saturn fits well within the center circle of our soccer field, as long as we don't include its rings. With the rings—the visible rings, anyway—Saturn is the same size as Jupiter. Uranus's diameter is 11 feet 9 inches (3.6 meters), roughly equivalent to the length of a Mini Cooper sitting on the field. Beside it, Neptune is also a Mini Cooper, but minus a bumper; it's just barely 11 feet 6 inches (3.5 meters) in diameter. So far, we've had a gander at the largest bodies in the solar system. Now, Earth is next. Earth is the size of a comfy arm chair, 35 inches (90 centimeters) across; followed by Venus, just about the same size at 33 inches (86 centimeters)—then Mars, the size of a drinking fountain—19 inches (50 centimeters) across. On our soccer field, Mercury is as wide as a laptop with a 15-inch (35-centimeter) screen. At this scale, the moon's diameter is the same as the length of a 10-inch (25-centimeter) tablet. You may be thinking: He must have exaggerated about the proportions, because the differences don't seem so big—it wasn't that bad at all. And you're right. It wasn't that bad, but only because I didn't mention distances. At this soccer field scale, the distance between the Sun and Neptune is about the same as the distance between Philadelphia and Boston. And, still at the same scale, the Oort cloud is pretty much as far out as the moon.

So, let's talk about distances. Please calmly forget about the sizes of the various bodies in the solar system. We'll put the Oort cloud aside for the moment, because it is one thousand times farther from the sun than anything else in the solar system; it's at a completely different scale. Let's take the distance from Earth to the moon as our standard unit of measure. It's a distance we can understand, and humans have traveled that distance multiple times. The distance between Earth and the moon is 220,000 miles. We'll treat that distance like a meter stick. Put a blue ball at one end—that's Earth—and a gray ball at the other end—that's the moon. The sun is 500 meters away from Earth—and at this scale, the sun is over 9 meters in diameter. Let's start at the sun, and from there, we'll travel through the whole solar system at this scale. After 190 meters, we meet up with Mercury; go 170 meters and we cross paths with Venus, where we are now 360 meters from the sun. At 500 meters, we go by Earth, and 260 meters farther out, we arrive on Mars. At 1 kilometer (0.6 miles) from the Sun—500 meters from Earth—we start evading asteroids, because we are now in the asteroid belt; we have to travel another 750 meters of our Earth-moon meters—about the same as the distance between the sun and Mars—before we leave the asteroid belt. At 2,600 meters from the sun, we find Jupiter; at this scale it measures approximately 90 centimeters (36 inches), and its red spot is around 10 centimeters (4 inches). Go another 2,150 meters farther, and we've finally reached Saturn and its rings. We are now 4,750 meters (3 miles) from the sun. Double the distance and we have almost arrived at

Uranus, which is found 9,600 meters from the sun. About 5 kilometers farther, at 15,000 meters (9 miles) from the sun, we reach Neptune, and at just about the same time, we enter the Kuiper Belt. By this point, if we had put the sun at the base of the Eiffel tower, we would now be in the gardens at the Chateau Versailles. Lovely. Another 12,500 meters farther out, we finally leave the Kuiper Belt. We have to travel another 25,000 kilometers to reach the Oort cloud, which is about equivalent to being in the middle of Australia, but taking the long way around. There are no two points on Earth that are 25,000 kilometers apart. Think about it—once two places are 20,000 kilometers apart, there's always a shorter route—by going in the other direction.

And voila! That's the solar system we live in. One of our closest neighbors is Proxima Centauri, a planet found at six times the distance between the sun and the Oort cloud—using an Earth-moon meter as a meter, the distance to our neighbor is almost half the actual distance between the earth and the moon. It's within the realm of possibility that the two systems have been continuously sharing matter, which makes Proxima Centauri a bit more than just a neighbor. And Proxima Centauri is only one single star among hundreds of billions in our galaxy, which is only one single galaxy among . . . I digress. It's still a little too soon for you to feel the dizzying vertigo of the immense distances we're talking about.

After this tour of the solar system, it's worthwhile to get an understanding of how it all stays in place, because, after all, why do planets have to revolve around a star? Why doesn't it all collapse in on itself? You'll see that the answers are incredibly simple, given the complexity of the results. These answers are false, of course, but they are often close enough to the right answer to be useful. Simple answers are a necessary step toward a better—more complicated,[83] yes, but also more correct—understanding of how bodies move with respect to each other. The branch of science that explains the motion of bodies is called mechanics. And the version we're interested in right now is classical mechanics.

---

[83]   Oh yeah, way more complicated! But don't be afraid, we're going to get there gradually.

CLASSICAL MECHANICS

Even the story about the apple isn't true.

Let's be clear on two things, and once they've been said, they are obvious: First of all, *classical* mechanics—no, not people who wear togas while fixing chariots—wasn't called *classical* until the theory of relativity came along. While it was being established, classical mechanics was simply called *the laws of mechanics*. Second, *mechanics*—classical or not—is the branch of science concerned with the motion of bodies; *bodies* is scientific lingo for any object at all: a car, a pebble, a planet, you, me, my hair—all of these are bodies.

It doesn't have much to do with what we usually think when we hear the word *mechanic*—someone who can fix a car. But the origins of the words are the same, and the earliest study of mechanics was as much about understanding motion as about being able to build machines that transform motion. Oh, and one last thing: Classical mechanics is often called *Newtonian mechanics*, even though they aren't exactly the same thing. But before we even think about discussing Newton . . . you guessed it: Aristotle.

43. The Great Question of Life, the Universe, and Everything

In *The Hitchhiker's Guide to the Galaxy*, researchers from a hyperintelligent, pan-dimensional species built the second greatest computer of all time, Deep Thought, to calculate the answer to the ultimate question of Life, the Universe, and Everything. After seven and a half million years of thinking about the question, Deep Thought finally came up with the answer: forty-two.

"Forty-two!" yelled Loonquawl. "Is that all you've got to show for seven and a half million years' work?"

"I checked it very thoroughly," said the computer, "and that quite definitely is the answer. I think the problem, to be quite honest with you, is that you've never actually known what the question is."[84]

44. Aristotle and Impetus

Even in prehistoric times, humans noticed the regularity of celestial motion;[85] this led them to envision some cosmic mechanism responsible for this order. But it took until antiquity for people—the species, not the magazine—to take a closer interest in what makes bodies move. Our buddy Aristotle, who was interested in just about everything, wondered about the nature of motion. He could see that there were two different kinds of motion, which could have been called natural and artificial, but which he instead called natural motion and violent motion.

To Aristotle, natural motion was motion that tried to steer a body toward its natural place. Remember, to Aristotle, matter is made up of five elements: water, earth, air, fire, and for everything between, ether. So, when you throw a stone, it naturally seeks to return to its natural place, which is earth. A flame, on the other hand, rises to reach its natural place, which is air. Even though the concept of five elements is a really moldy old theory of matter, we should give Aristotle credit for at least trying to explain, by way of this theory, what would today be a simple observation.

Next is violent motion. Violent motion was everything that isn't a natural motion. For example, the act of throwing a stone makes it travel in a direction it wouldn't have taken by itself: This is a violent motion. Aristotle thought that violent motion could happen only in the presence of a force. Basically, his intuition and actions clearly showed him that a stone,

---

[84] Douglas Adams, *Hitchhiker's Guide to the Galaxy*, vol 1. And Douglas Adams, thanks for all the fish.

[85] In Scotland, archaeologists discovered a lunar calendar that is 10,000 years old.

once you stop bothering it, didn't just fly off by itself. As long as no force was applied to it, it stayed where it was or it followed its natural motion.

We should give Aristotle credit here—and this is the last time I'll give him credit for anything; I promise—for the fact that he had issues with his own explanation. According to his explanation, when you throw a stone in the air, and you are no longer in contact with the stone, no more force is applied to it, and it should immediately stop its violent motion and simply drop straight down—which is its natural motion. Yet the stone continues to follows its path for a certain period of time. In the absence of any better explanation, Aristotle explained that as the stone moved forward, it left a vacuum behind it that was immediately filled with air—which was air's natural motion—and so it was air that continued to apply a violent force to the stone in question, allowing it to follow its path for a few meters.

This problem is known as the "arrow problem": when an archer shoots an arrow, the moment it leaves the bow, no more force is applied to it. But a well-shot arrow doesn't travel just a few meters, it travels dozens of meters. And yet, for lack of any better answer, Aristotle's idea persisted and dominated how the Western world thought about motion until the beginning of the sixth century—that is, almost a thousand years. In 517, the grammarian and philosopher John Philoponus of Alexandria was the first to explore—in a written document, anyway—a different idea: the idea that a projectile continues to move forward not because it's pushed by the air, but because of a motive force transmitted by the projector at the moment the projectile is launched. This theory is known by the Latin term *impetus*.[86]

The idea of impetus is this: When an archer shoots an arrow, a force, or perhaps more precisely a reserve of force, is transmitted at the moment of release, which keeps the arrow in motion. How quickly the reserve runs out depends on the amount of force transmitted, because air works against the motion of the projectile—contradicting Aristotle's original ideas. Once this reserve runs out, the arrow is no longer driven by anything other than its natural motion toward the earth.

[86]   Meaning "attack."

The impetus idea spread, and until about the year 1500, folks thought that when a projectile was launched, it followed a straight trajectory as long as it had impetus, and then it followed another trajectory—still straight, but now vertical—as soon as it started its natural motion. People—the species, not the magazine—then started talking about violent impetus and natural impetus, because they had nothing else to explain what causes the natural motion of a body. Little by little, by pondering the problem, plus some intuition, and maybe just a little observation, an idea sprouted: The violent impetus transmitted to the arrow by the archer's bow was gradually transformed, as a result of the air's resistance to the motion, into natural impetus. That meant, at the beginning of the seventeenth century—finally!—the curved trajectory of a projectile was explained. It wasn't completely right, of course, but the theory was undeniably the starting point for the concept of inertia. This important step allowed a more serious theory of mechanics to develop.

45. Archimedes and the First Mechanics

Before being a branch of physics, mechanics was at first a branch of mathematics. Until the eighteenth century, the natural field of application for mathematics was mechanics, and problems in mechanics were naturally expressed as geometry problems, and more generally, as mathematical problems.

Because of this overlap, the most brilliant mathematician of antiquity, Archimedes of Syracuse—born more than fifty years after Aristotle's death and a long time before any talk of impetus—became the father of statics. Statics is the branch of mechanics concerned with equilibrium—that is, what happens when no force is acting or when a set of forces cancel each other out.

While he didn't invent these concepts, Archimedes studied the lever, the endless screw, gears—basically, what we would generally call today simple machines. He showed that with an ingenious system of pulleys, a man could lift objects much heavier than his own weight.

From there, Archimedes went on to invent various machines using compound pulleys. Using the law of the lever, he showed it was possible to amplify movement significantly. And he invented new military technologies: simple ones like the arrow slit and more complicated ones like the catapult.[87]

If the distinction had existed at the time, Archimedes would have been called an inventor rather than a scientist. For him, understanding why a catapult transmitted a force was more about practical solutions than theoretical questions. Today, we would say that he was blessed with amazing instincts and an extremely ingenious mind. The principle that bears his name is a typical example.

Archimedes's Principle

A body submerged in water comes out wet. . . . No, that's not it.

In his treatise *On Floating Bodies*, Archimedes was concerned with the branch of mechanics now called hydrostatics. In it he expressed this famous principle, which goes as follows:

Any object, wholly or partially immersed in a fluid, is buoyed up by a force equal to the weight of the fluid displaced by the object.

In other words, if you immerse a body in water, it displaces a certain amount of water. This amount of water has a volume. The immersed body experiences a force that opposes its downward motion—an upward force trying to make it go back up again. The magnitude of this force equals the weight of the volume of the displaced water.

The only reason it's called Archimedes's *principle* is that Archimedes's only proof of this effect was the fact that his experiment worked.[88] The reason it is now called the *law of buoyancy* is because the underlying theory has been successfully laid out and verified for quite some time now.

---

[87]   Used in the cow-tossing scene in *Monty Python and the Holy Grail*.

[88]   In the physical sciences, a *principle* is a law that is accepted but that has never been demonstrated theoretically.

The principle allowed him to invent, and have Archias of Corinth build, the largest transport ship of antiquity, the *Syracusia*, by order of Hiero II, tyrant of Syracuse. Hiero II and Archimedes have other bits of history in common, particularly a story about a certain crown.

46. Eureka, or the Golden Crown of Hiero II, Tyrant of Syracuse

Marcus Vitruvius Pollio, an architect in the first century BCE—born more than 120 years after Archimedes's death—tells us this famous story:

Hiero II, tyrant of Syracuse, gives a goldsmith some gold to make a solid gold crown. But Hiero isn't the trusting type, and he worries that the goldsmith might have swindled him by making the crown out of a cheaper metal, like silver, then gilding it, then keeping the leftover gold for himself. But how can he know for sure? He doesn't want to risk cutting or melting the crown only to find out it was really solid gold after all.

So he asks his friend Archimedes to find a solution to the problem and confirm whether his crown is made of gold or not, but without damaging it. If it turns out that the goldsmith has ripped him off, the goldsmith will certainly pay with his life. Even though Archimedes's own life wasn't at stake, it's always a good idea to keep your tyrant happy if you want to continue working on your science in peace.

The solution comes quickly, but the implementation poses a problem. Basically, Archimedes knows that if he can figure out the density[89] of the crown, he would know right away if it has the same density as solid gold. He starts by trying to model the crown geometrically, but it contains too many details, angles, curves, points, hollows, and reliefs to measure its volume precisely. With only an approximate idea of the crown's volume, he wasn't sure that it contained only gold. Now what?

One day, as Archimedes gets into the tub at the public baths, he notices that the water level rises slightly—and something clicks in his brilliant brain. He realizes the connection between a body immersed in

---

[89] Density is the amount of matter or mass per unit of volume. The expression *volumetric mass density* is also used.

water and the effect it produces! He is on the verge of discovering his famous principle. Convinced that he has the first part of the solution, he jumps out of the bathtub and rushes home, running naked through the streets shouting "ηὕρηκα"![90]

Back at home, he starts doing experiments to figure out, quantitatively, exactly what happens when a body is immersed in water.

He ends up discovering the link between the displaced volume and the density of the immersed body. He finally has a way to measure the density of the crown without damaging it.

He put his solution into practice: First, he weighed the crown with a balance scale; then, he cast a pure gold ingot that weighs the same as the crown. Since the two objects are the same weight, they are assumed to contain the same quantity of gold and would, therefore, have the same density. He then weighs the crown again using the balance scale, but this time the crown is completely submerged in water; then he does the same thing with the ingot. He can then compare their density and work out whether the crown is made of pure solid gold.

It isn't. The crown is definitely made of some gold, but also of another metal, most likely silver. Hiero is both furious with the goldsmith—goldsmith: executed—and ecstatic about Archimedes's work—Archimedes: not executed.

This story is so legendary that it may in fact be only a legend: Vitruvius is the only person who tells this story, even though he wasn't a contemporary of either Archimedes or Hiero II. The consensus among historians today is that while Archimedes probably wouldn't have had any trouble solving such a problem, the anecdote itself is probably an ancient urban myth. And that's a shame, because it's such a great story.

47. Good Old Science Dude: Galileo, Part 1

Galileo Galilei—oh, that name! Sorry if I sound like a broken record here, but imagine if the most brilliant scientist today was named Albert Talbert—ahem. Galileo Galilei can be considered the father of modern

---

[90] *Eureka*, in Greek: "I've found it!"

science. He not only gave us the basis of classical mechanics but he also set up a procedure we know of today as the *scientific method*.

Born in Pisa, Italy, in 1564, Galileo was the eldest of seven children. At the Santa Maria de Vallombrosa Abbey, his instructors expected him to enter religious life; however, that idea didn't exactly thrill him. Early on, Galileo's father realized that his son had astonishing intelligence. The fact that young Galileo had an eye disease enabled his father to enroll him in the University of Pisa to study medicine, which he also wasn't interested in. His father sent him to the university hoping he would follow in the footsteps of his ancestor, Galilaeus Galilaeis. Seriously folks, there are other first names you can use.

At the university, Galileo was introduced to mathematics by Ostilio Ricci. This was indeed fortunate, because Ricci had the spectacular habit of matching theory with practical experience (i.e., reality), particularly through experiments. In 1584, Galileo measured the swinging of the chandeliers in the Cathedral of Pisa; he used his own pulse to time them—stopwatches weren't a thing yet—and he was only nineteen years old at the time! He was working on the isochronism of pendulums (we'll get to that explanation in about five seconds) seventy-five years before Huygens, the Dutch mathematician who found the mathematical formula for isochronism.

Isochronism of Cycloidal Pendulums

Isochronism of cycloidal pendulums, as shown by Huygens, is the phenomenon that when these pendulums swing, the period (the time it takes for a pendulum to make a complete trip back and forth) is the same—regardless of the amplitude (how far the mass travels out and back). We don't need to go into detail here about cycloidal pendulums; they are just regular ol' pendulums: a string attached to a fixed point at one end and a mass hanging from the other end, with the mass following a specific curved path as it travels.

Anyway, the work Galileo did on the subject provided the first step into a completely new territory of understanding: Galilean mechanics. With his observations and experiments, Galileo showed us that the period of a simple pendulum depends—ha! de-*pend*-s—only on the length of the string. That's the law of isochronism.

Basically, by 1585, Galileo had had enough of the university. When it came to Aristotle and medicine, Galileo was not a fan. Galileo abandoned his studies at the university and returned to Florence without a degree. For a while, he immersed himself in the intellectual circles in Florence, which included the musicians and artists who had influenced his father. Galileo, mind blown by the works of Euclid, would indeed follow in his ancestor's footsteps. He would also stir up renewed interest in the ideas of Plato, Pythagoras, and Archimedes. Galileo was also beginning to seriously doubt Aristotle's geocentric ideas about the solar system and the universe.

Earth Is the Center of the Universe, by Aristotle

Aristotle thought that the earth was fixed—stuck in one place, doesn't move. Period.

Aristotle was *that kind of guy* . . . he just had a look around and saw it was obvious that the earth wasn't moving. Now, all he had to do was prove it. Aristotle's proof of the immobility of the earth shows (1) how he liked to prove ideas just by talking about them, and (2) how his arguments contained the conclusion in their initial assertion. Aristotle said the following: If you drop an object (*object* means "a thing"—scientists like to use the word *object*, because it has more syllables than *thing*, so they feel smarter), it falls vertically— straight down—to the ground. If Earth were moving, it would move while the object was falling, thus you would see that it isn't falling straight down.

So, if the earth doesn't move—that is, is fixed—and we can plainly see the sun, the moon, and the stars are all moving, it's because *they* go around the earth. Voila! There you have it: Proof that the earth is the center of the universe. Giordano Bruno will have the pleasure of putting Aristotle in his place.

Thanks to a recommendation, Galileo was named chair of mathematics at the University of Pisa in 1589. Thus he returned to Pisa, this time with mathematics as his top priority. There, Galileo tackled one of

Aristotle's ideas that he had a problem with: the idea that heavy objects fall faster than light ones.

Galileo argued as follows: If I drop a heavy ball of lead and a light ball of cork from the same height, the lead ball—being heavier—will fall faster. OK, now, what happens if I tie them together? The lead ball should fall faster and pull down on the cork ball, making it fall faster. But wait, the cork ball falls more slowly, right? Does it hold back the lead ball and slow it down, making it fall more slowly? How can the two pieces tied together fall faster and slower at the same time? They can't. The only possible answer is, together or apart, the two objects must fall at the same speed.

That said, however, general observation appeared to show that Aristotle was right. Everyone knows full well that a heavy object falls faster than a light object. Galileo thought that the air prevented the light objects from falling as fast as the heavy ones. He must have had Archimedes's principle stuck in some corner of his brain. Galileo wanted to prove his hypothesis. But how? He would do something revolutionary— he would do *experiments*!

In order to minimize the effects of the air (i.e., air resistance), he would use objects with the same shape but not the same mass. He used a ball made of lead and a ball made of cork, both the same size. He asked assistants to drop both items from a window simultaneously to see if they did, or didn't, hit the ground at the same time. And they certainly did. He repeated the experiment over and over using larger and smaller items, always both the same size as each other. And over and over, he got the same result: The speed of a falling object did not—and still doesn't—depend on its weight. Sorry, Aristotle, but nice try.

Today, most historians agree that Galileo never actually performed the experiment. A few holdouts wonder whether he did it at Pisa, or maybe in Padua; still others think he did it at home in Florence. Hard to say . . . and is it really all that important anyway? Even if he never performed the experiment at all, what *is* important here is that he had enough confidence in his own intuition (and in the paradox he found with the two objects tied together) to refute commonly held ideas.

Let's take a little detour here to ask ourselves the same question. After all, it is a fact that when an object has a greater mass, its weight (i.e., the force that attracts it toward the ground) is indeed larger. It seems natural to think it will fall faster. But it doesn't. How do we explain that? It's because of a property of physical bodies called inertia. It's a concept we haven't really discussed yet, but it's absolutely essential for the understanding of the laws of mechanics.

Inertia

Inertia is a property of a physical body, just like mass is. Very often, inertia is used synonymously with mass—but we'll talk about that later. Inertia can be defined as a body's resistance to starting to move when it's standing still (as the experts say, "at rest") and as a body's resistance to changing its motion when it's already moving. If you stop and think about it, it's obvious. Imagine a soccer ball and a car with free-rolling tires—no brakes—and both of them are on pavement that's perfectly flat. Go ahead, push the ball. Now, push the car. You can see that it's much easier to get the ball rolling than it is to get the car rolling. And once you get them moving, it will be also be easier to stop the ball than it will be to stop the car. Our intuition tells us: It's because the car is heavier, because the mass of the car is greater. In a way, that is correct. But truth be told, it's because the inertia of the car is greater than the inertia of the soccer ball. And inertia is directly (totally, completely) proportional to mass.

When you drop two objects from the same height—imagine we are doing it in a vacuum, so there's no air resistance— let's assume that one of these objects is a wrecking ball, weighing several hundred pounds, and the other is a Ping-Pong ball. The wrecking ball has so much inertia that a huge amount of force is needed just to get it moving. Just because it falls naturally due to gravity, it doesn't change anything—everything happens as though part of the force of gravity were used "just" to get the ball moving, which is wrong, but it gives you the idea. When it comes to the Ping-Pong ball, significantly less force is needed to make it fall because its mass is so much smaller. As it turns out, inertia is proportional to mass, so the difference in force needed exactly corresponds to the difference in mass.

So, the more an object weighs, the more difficult it is to make it fall, these two "mores" being of equal magnitude. This means extra inertia offsets extra weight, and an object, whatever its mass, will always fall at the same speed—as long as air resistance doesn't come into play. End of detour.

In 1592, Galileo went to Padua to teach at the university. He stayed there for eighteen years. Padua is in the Republic of Venice and was just about outside the reach of the Inquisition; thus Galileo enjoyed a great deal of academic freedom in Padua. He set up partnerships with the local smelters and carpenters, which let him develop various experiments with his students. Though he had personally adopted Copernicus's heliocentric model, he wisely continued to teach the Ptolemaic model until there was sufficient proof to be rid of, once and for all, the Aristotelian model of the earth at the center of the universe and celestial vaults made of clear crystal. He taught astronomy, mechanics, and mathematics. At the same time, he ran up enormous debts after the death of his father in 1591, when he became the family provider. He improved various measuring instruments and military equipment and kept teaching until his *annus mirabilis*,[91] in 1604.

In July, he perfected a water pump that he installed in the gardens of Padua. He discovered years later that it's impossible for a suction pump to pump water from more than ten meters deep, yet he couldn't explain why; but that's another story, which involves Evangelista Torricelli. In December, he observed a nova, which completely contradicts the immutability of the sky proposed by Aristotle. In public, however, Galileo remained Aristotelian, feeling he still didn't have enough proof—especially because the nova returned to its normal appearance.

Nova

Without going into details, a nova is a star that shines much brighter all of a sudden and then returns, a few days later, to its normal appearance. This can be a regular, recurring phenomenon. Or not.

[91] Miraculous year.

Actually, I am most interested in October 1604, because that's the month Galileo discovered the law of uniform acceleration, the outcome of his work on falling bodies. He showed that as long as objects are subject only to their own weight, and other forces such as air resistance are negligible, the fall of two objects under the same conditions does not depend on their mass, because their "inertial mass"—their inertia— exactly offsets their "gravitational mass"—what is traditionally called mass—namely, the quantity of matter. To demonstrate this, Galileo organized "slow-motion" falls, by sliding disks with different masses down iced inclined planes—kind of like our ball rolling on a slanted board experiment (see page 4)—the ice minimized the friction between the disks and the surface.

After 1604, Galileo entered his great astronomical period. In 1609, he received a letter from Paris from one of his students, telling him that the Dutch optician Hans Lippershey had developed a telescope that made it possible to see distant objects, magnifying them by up to seven times. Even though Lippershey's telescopes were fairly ineffective due to multiple optical aberrations (weird effects), Galileo saw the potential and built his first telescope. In 1609, he finished the second version of his telescope and presented it to the Venetian senate. And mamma mia! They were so excited about the clarity of the image and the fact that the island of Murano and many other things seemed eight times closer than they really were.

Galileo offered the instrument and his inventor's royalties to the Republic of Venice. The Republic of Venice then doubled his salary and guaranteed his university position for life. Finally! This gave him some financial breathing room. He then improved the design to magnify objects by more than thirty times and cleverly used a concave lens to correct the aberrations. But he wasn't an engineer at heart, so the telescopes he made were only sort of effective. Though the design was excellent, the execution left something to be desired.

So Galileo set about watching the heavens and immediately found a ton of new evidence to challenge Aristotle's guidelines. Because basically, according to Aristotle, there were two distinct worlds: the *sublunary world*, which includes the earth and everything up to the moon, and the *superlunary world*, which includes everything beyond, which was necessarily geometrically perfect—spherical—immutable and regular in its motions. Galileo looked at the moon and observed that the terminator—not Arnold Schwarzenegger, but rather the boundary line between the illuminated day side and dark night side of the moon—was irregular, and that the moon's surface was studded with mountains. A few weeks later, he discovered the nature of the Milky Way and sun spots, he discovered some stars are actually clusters of stars, and he found that Jupiter has moons. To flatter his patrons, the Medici family, he called these moons the *Medicea sidera* (Medician stars).

The discovery of those moons was a real problem for the Aristotelians, who believed everything went around the earth. The fact that these moons moved with Jupiter and traveled around it seemed to contradict that principle. Galileo readily saw the discovery as a "model" of the solar system, as Copernicus had conceived it—bar one difference: namely, that Copernicus thought everything except the moon revolved around the sun. When Galileo published his results about the moons, copies of the *Sidereal Messenger*[92] sold out in just a few days.

It was definitely a hit. Even though he didn't have a refracting telescope, Johannes Kepler, the imperial German astronomer, lent his support to the discovery in a work called *Discussion with the Sidereal Messenger*.[93] He commented on the remarkable impact of such a discovery and speculates about its consequences for astrology. In September 1610, he published a *narratio* recounting his own observation of the Jovian companions and coined the word *satellite*, which means "bodyguard" in Latin. It's interesting that Kepler had his own theories about the solar system; it was supposed to be a system of five solids and could not contain more than six planets. Kepler was obsessed with astrology—you just never know about some people—and Galileo deplored astrology, perhaps to a fault; it's the reason Galileo openly ridiculed Kepler's idea that the moon influenced the earth's tides.

[92] *Sidereus Nuncius.* FYI: *sidereal* means "of, or relating to stars."
[93] *Dissertatio cum Nuncio Sidereo.*

Next, Galileo discovered Saturn's rings, but he didn't understand them (see page 137). Then he discovered the different phases of Venus—another harsh blow for the Aristotelians, because this phenomenon is fairly inexplicable for a geocentrist. Galileo put himself more and more at risk because, while we haven't mentioned it yet, there was a much more significant problem than the impact on Aristotle, who had been dead for a very long time and who was constantly turning over in his grave because of the things proclaimed in his name. The problem was Psalm 93: "Indeed, the world is established; it cannot be moved."[94] The Church began to take a more serious interest in the ideas of this Galileo, who was certainly enjoying more and more success with his heliocentric theories. Galileo knew very well that he needed to be careful, because a name still reverberated in the back of his mind: Bruno. Giordano Bruno.

48. Giordano Bruno, Punk Genius and Father of Relativity

Galileo was not the first person to question the quality of Aristotle's theories. You will see that Giordano Bruno was less—much less—cautious than Galileo in taking those theories apart one by one. Remember how Aristotle demonstrated that the earth was immobile, and how it followed that everything else must revolve around the earth, because everything else was revolving? Aristotle argued that if you drop a stone from the top of a tree, it falls straight down, landing directly below where it was dropped, thus proving the earth is immobile, because if the earth turned as the stone fell, the stone wouldn't fall straight down. So, Giordano Bruno said to himself: "Well, in that case, if I drop a stone from the top of the mast of a moving boat, then the stone shouldn't fall straight down," and Bruno decided to try the experiment.

So he dropped a stone from the top of the mast of a moving boat and he saw that it fell straight down. Whether the boat was moving relative to the riverbank or not—that is, whether the boat was sailing down the river or just tied up at dock, the stone fell straight down. Here's his conclusion:

94   . רְאא-זָוכָּת לֵבַת,לבַּ-טוֹמַת

> All things located on Earth move with the earth. A stone thrown
> from the top of a mast will come down, moving as the boat moves.[95]

Bruno had discovered that no motion can be described as absolute, and that the boat was fixed to those who were on it but moving from the perspective of someone on the riverbank. What he had discovered: Motion is always relative.

Based on this experiment, Bruno was able to take apart Aristotle's first argument, then unravel the whole thread of arguments, and there was no longer any reason to believe everything revolves around the earth. With that, Bruno provided the first glimmer of relativity,[96] leaving a deep and satisfying boot print right on Aristotle's face.

Bruno stayed in Italy as long as he could, but an inquisition for heresy was opened against him in 1576—perhaps I should mention that eleven years earlier he had become a Dominican priest, but then he eventually rejected their dogma, got interested in magic and alchemy . . . and, in short, did all kinds of things that would get him kicked out— and he had to flee. He wandered from town to town until, in Geneva, Switzerland, he joined an evangelical community. They excommunicated him in 1578. He went to Lyon, France, then south to Toulouse, also in France, where he authored *Clavis Magna*, a book about mnemonic devices. King Henri III of France was so impressed that he summoned Bruno to his court and offered to protect him. Bruno could breathe easy.

From that point, Bruno goes further than anyone else before him. He quickly adopts Copernicus's heliocentric model, but that isn't enough. He couldn't see why the planets revolving around the sun meant the solar system should have any special privilege. Bruno put an end to the idea of a finite universe by pointing out that, in his opinion, it extended in all directions to infinity and was filled with many stars. Bruno had just blown the walls off space, and people—the species, not the magazine—started talking about the universe. And he continued, if

---

[95] Giordano Bruno, *The Ash Wednesday Supper*, 1584.

[96] Special relativity, that is.

the sun is a star like the others, then there must be planets orbiting them, and if among the planets in our system, there is life on a planet, Earth, then there must be life on other planets in other star systems. To Giordano Bruno, the universe was made up of an infinite number of stars orbited by planets that could potentially harbor life. He assumed there was life elsewhere in the universe: "We declare that there is an infinity of earths, an infinity of suns, and an infinite ether."[97]

And there's more:

> Because it is impossible for a rational, sufficiently vigilant person
> to imagine that these innumerable worlds, as magnificent as or
> even more wondrous than our own, would be without similar or
> even superior inhabitants.[98]

Bruno presented his cosmogonic[99] vision all over Europe, with varying degrees of success—in England, at Oxford, his vision ruffled the feathers of the Church of England—and in 1585 he started working on *Figuratio Aristotelici Physici Auditus*,[100] a detailed critique of Aristotle's scientific ideas. Due to politico-religious reasons, Henri III couldn't afford to protect Bruno any longer. So Bruno left France in 1586 and went to live in self-exile in Germany. There, he joined the Lutheran community. And it should come as no great surprise that in 1588 Bruno was excommunicated from the Lutheran Church.

In 1591, Giovanni Mocenigo invited him to come back to Venice— remember, the Inquisition had very little power there due to the influence of the Republic of Venice—and Bruno, who was very homesick, agreed to return. Unfortunately, there was a bit of a misunderstanding: Bruno was hoping to be appointed as chair of mathematics at the University of Padua— this is one year before Galileo started to teach there—and Mocenigo was expecting one-on-one lessons from Bruno in mnemonic techniques.

---

[97]   Giordano Bruno, *On the Infinite Universe and Worlds*, 1584.
[98]   Ibid.
[99]   Wow! What a great word for "a theory or model of the universe."
[100] Giordano Bruno, *A Statement of Aristotle's Physics*, 1585.

Mocenigo felt like he wasn't getting his money's worth, and Bruno felt underappreciated—wasn't it honor enough that he had accepted Mocenigo's invitation? When Bruno tried to leave, Mocenigo held him against his will, but he couldn't get Bruno to acquiesce—many others had already run into that brick wall. On May 23, 1592, Mocenigo denounced Bruno to the Venetian Inquisition. And that was the beginning of the end.

Although Bruno was exonerated by the Venetian authorities, Pope Clement VIII commanded the doge to extradite Bruno, and it was an offer he couldn't refuse. For the rest of his life, Bruno was a prisoner of the Holy See under the jurisdiction of the Roman Inquisition. Bruno's trial lasted for eight years, and the charges against him evolved over time. First, his theological positions were described as heretical—rejecting the tenets of the Church, such as the Trinity, Mary's virginity, and transubstantiation. Then, there were his more or less occult activities—alchemy, astrology, divine magic, transmigration of the soul (the ability to transfer his soul into another body)—and finally, his cosmogonic vision. The months passed and his rap sheet grew longer and longer.

The trial revealed the truly punk soul of Giordano Bruno, because Bruno was unbreakable. For a long time he had the option to remedy the situation: to recant, and accept the teachings of the Church and admit his errors. He wouldn't do it. Inquisitor after inquisitor threw in the towel. No one seemed to be able to wear him down. He wrote retraction statements with some of them, dictating the wording, forcing them to make additions, and corrections, and revisions—remember, folks, no word processors, no copy-paste, no Wite-Out, all handwritten changes. Each iteration took several months—in the meantime, he was tortured and starved—but the inquisitors wanted to wrap it up. Once the retraction document was finalized, with the exact wording Bruno wanted, with all the changes he had approved one by one over many long months and years, when all he had to do was simply sign the papers and save his skin, he did a sudden about-face and completely rejected the whole thing: "I do not shrink from death and my heart will not submit to any mortal."

Even Pope Clement VIII gave it a try. And one last time, he asked Bruno to submit. Bruno's reply: "I fear nothing and I retract nothing, there is nothing to retract and I don't know what I would retract."

This time, it was just too much—it had gone on for far too long. Clement VIII ordered the tribunal of the Inquisition to pronounce its judgment, and it sentenced Bruno to be burned at the stake on January 20, 1600. At his sentencing, Bruno declared to his judge: "Perchance you who pronounce my sentence are in greater fear than I who receive it."

Bruno was executed on February 17, 1600, burned at the stake in the Campo de' Fiori square in Rome. Today, you'll find a statue there, erected in his honor. The Vatican has never revised this ruling. In fact, in the twentieth century, Pope Pius XI beatified, then canonized, and finally declared Cardinal Robert Bellarmine to be a Doctor of the Church. This is the same guy who headed the investigation for Bruno's trial. This confirmed the Church's position on the subject in a rather final way. Pope John Paul II reaffirmed this position when he revised the Church's position on Galileo in 1981.

As you can imagine, if Galileo valued his skin, he'd have to be more careful than Bruno had been.

49. Good Old Science Dude: Galileo, Part 2

Previously, in "Good Old Science Dude" . . . Galileo had attacked Aristotelian theories harshly and had ridiculed Aristotle's geocentric model. Galileo was under more frequent attack by those who supported a geocentric model. And here's the scary part: more of the attackers were Church scholars. Galileo was familiar with Giordano Bruno's theories. He was also well aware what they had cost him. Galileo realized he'd have to be much more careful.[101]

Since Galileo was obviously a grand master at astronomy, his opponents attacked him regarding another area of study: floating bodies. Galileo said that ice floated on water because it was less dense, while Aristotle claimed it was ice's nature to float. Galileo proved the truth of

[101]   If you really can't remember the story, it's on page 166.

his statements and emerged victorious in what came to be known as the "Battle of the Floating Bodies."

Then, in 1612, a German Jesuit astronomer, Christoph Scheiner, attacked Galileo's theories on sun spots. According to Scheiner, the sun was perfect and incorruptible by nature and, therefore, could not have spots. So sun spots could only be clusters of stars located between the earth and the sun. In response, Galileo brilliantly demonstrated that the spots could only be on the surface of the sun, or at least so very close to the surface that it wasn't possible to determine an exact altitude. Their letters to each other were published in 1613 under the title *History and Demonstrations Concerning Sunspots and Their Properties*,[102] or *Letters on Sunspots* for short—Scheiner eventually came to agree with Galileo's theories, and we really should acknowledge that as a very noble gesture.

In November 1612, the Church strengthened its stance on both the heliocentric model and the earth's rotation. Niccolò Lorini, a Dominican priest and professor of ecclesiastical history in Florence, wrote a scathing discourse that vehemently attacked the theories. He specifically cited the scriptures to prove that the sun and the moon revolve around the earth:

> Then spake Joshua to the Lord in the day when the Lord delivered up the Amorites before the children of Israel, and he said in the sight of Israel, *Sun, stand thou still upon Gibeon; and thou, Moon, in the valley of Ajalon.*[103]

He argued that if it's possible to tell the sun and the moon to stop moving, it's because those two heavenly bodies are moving, thus proving that the sun and the moon revolve around the earth.

On January 6, 1615, Paolo Antonio Foscarini, a Copernican, published a letter in which he not only defended Copernican heliocentric theories but actually promoted them as a *physical reality*! This was so

---

[102]   *Storia e dimostrazioni intorno alle macchie solari e loro accidenti.*

[103]   Joshua 10:12, King James Version; emphasis added.

scandalous that Cardinal Bellarmine—yes, the same guy who headed the investigation for Giordano's trial and also happened to be friends with Galileo—had to intervene. Bellarmine wrote a letter to Foscarini that commanded him to cut the crap big time—pardon my French—unless he could present irrefutable proof to confirm the heliocentric theory or at least disprove the geocentric theory.

Let's spend a paragraph or two, here, to highlight some important and possibly enlightening information. First, regardless of what you might think about the Church in the Middle Ages, it was actually open to new ideas. Although the Church had a rule that the holy scriptures were always just and true, the Church was allowed to admit that some *interpretations* could be wrong, but that's all. This partly explains why Bellarmine didn't just have Foscarini hauled off to the Inquisition. Because if there was irrefutable proof of the heliocentric model, the Church would admit that some interpretations of the scriptures needed to be reviewed, and leave it at that.

The second thing to know is that the Church was also open to new ideas from a practical standpoint. It had been widely known for quite a while that the heliocentric system greatly simplified calculating the positions of stars compared to using the geocentric model. But there's a key point here: Until proven otherwise, the heliocentric model was nothing more than a useful *model* that made calculations easier. That's why Foscarini's letter caused such a scandal; he claimed it was reality.

Galileo got involved in April 1615 by writing a long letter to Christine de Lorraine—the older sister of Charles III, Duke of Lorraine, and granddaughter of Catherine de' Medici. In his letter, he defended the orthodoxy of the Copernican model—he tried to show that the heliocentric model didn't contradict the scriptures. He wrote to defend that position. He also hoped to prevent a complete ban of the model. And he was absolutely outraged to have to deal with scientific debate sliding over into areas of faith.

Galileo went to Rome to try to prevent the theory from being banned; it was already under investigation, but he failed to demonstrate that the earth actually moves—he tried to prove it using the tides.

Although the Church offered to compromise, Galileo refused to consider those Aristotelian, geocentric theories of Ptolemy as equal to the Copernican theories. In February 1616, after spending months attempting a number of different compromises, the ban was approved by Pope Paul V and the Inquisition. The heliocentric model was declared to be contrary to the scriptures. They asked Galileo to not teach Copernican theories as other than unfounded, though practical, models. Because of a persistent rumor that the Inquisition had really put the screws to him, Galileo asked Bellarmine for a certificate stating that they had not put the screws to him. The certificate was given to him in May 1616. After that, things quieted down for a while.

In 1619, Father Orazio Grassi, a Jesuit priest, published a treatise on the elliptic motion of comets. Galileo rebutted it by pushing one of his own students to defend a crazy theory that came out of nowhere, saying comets were actually only optical illusions. Orazio Grassi wasn't fooled and he directly attacked Galileo in a cunningly dangerous pamphlet that cleverly mixed scientific considerations and religious insinuations.

Galileo's friend Cardinal Maffeo Barberini—the guy who would be the next pope, by the way—encouraged him to write an ironic response, *The Assayer*.[104] To this day, it is considered a "masterpiece in the controversial art of polemic essays."[105] Well, Grassi was humiliated, and he sent an anonymous letter to the Inquisition denouncing Galileo. The Inquisition ignored it. Galileo didn't worry about it. In 1623, Cardinal Barberini was elected Pope Urban VIII, and Galileo was permitted to publish *The Assayer*, which he dedicated to the pope. It was a hit. Even though Galileo wasn't seeking attention, he was suddenly the poster boy for intellectuals who were struggling with the Jesuits and their scientific theories.

In *The Assayer*, Galileo suggests that mathematics is the language of science:

---

[104]   Italian title: *Il Saggiatore*.

[105]   Galileo: *The Final Drama and the Crowning Achievement*, in Pierre Costabel, *Encyclopædia Universalis*.

The philosophy is written in this immense book, the universe. It's always wide open right in front of our eyes, but we won't understand it unless we first learn the language it's written in and understand the characters used to write it. It is written in the language of mathematics; its characters are triangles, circles, and other geometric shapes; without them, it's not humanly possible to understand a word of it.[106]

In the years that followed, Galileo continued to live his life despite failing eyesight and nearly continuous attacks from Aristotelians. They even tried to make him lose his stipend from the University of Padua, but they were unsuccessful.

After the Copernican models were banned, Galileo often went to Rome to visit his buddy, Pope Urban VIII. The pope asked him to present the pros and cons of the geocentric and heliocentric models in a book. He fervently wanted the book to present the models in a completely neutral, unbiased manner—Pope Urban VIII specifically reminded Galileo about the failed presentation on the tides and strongly suggested that he not present those kinds of ideas.

The book was printed in 1632; Galileo was sure his ideas would be an absolute triumph and would put an end to the ban! At least, that's what he wanted. Ah, the pride before the fall. The book was published with imprimatur, aka official approval from the Church. Galileo had gotten the approval by tricking Monsignor Riccardi. Galileo showed him only the foreword and the conclusion. Those two parts of the book were completely neutral, but the rest of the book openly ridiculed Ptolemaic and Aristotelian ideas.

The book, *Dialogue Concerning the Two Chief World Systems*,[107] was not his triumph, it was his undoing. In the book, he presented three characters talking about the nature of the world. Filippo Salviati, a Copernican; Giovan Francesco Sagredo, an intelligent, open-minded Venetian; and Simplicio, a defender of Aristotelian positions. The third

---

[106]   Galileo, *L'Essayeur* (*The Assayer*) (Paris: Belles lettres, 1980), p. 141.
[107]   Italian title: *Dialogo sopra i due massimi sistemi del mondo.*

guy is appropriately named: He is stupid, he asks only idiotic questions, and he has absolutely no idea how to defend his position other than on the basis of faith, which certainly irritated Church scholars.

A Boat Ride, Just Like Bruno

In the *Dialogue*, Galileo wrote about an experiment that looked a lot like the experiment Giordano Bruno had done on a boat, but Galileo's experiment would provide an additional and more important fact. He showed that if someone was inside the cabin of the boat, with no portholes or windows, and if they performed various mechanical experiments—like jumping in place or pouring a liquid, drop by drop, into a container placed directly below—it was impossible to tell if the boat was moving or not, as long as the boat didn't change speed or direction.

What Galileo said is, and this is key:

- First, this kind of motion—rectilinear, uniform motion—or no displacement at all: it's the same thing.
- Second, the laws of physics are the same whether you are moving or not.
- Finally, in this case, there isn't any experiment you can do to determine if you are moving or not, meaning there is no absolute motion; motion is always relative to something else.

So, right there, Galileo invented inertial reference frames and started paving the road to relativity.

Offended, the Church had to react for two reasons: First, they felt—and justifiably so—that Galileo had lied to get imprimatur by not clearly representing the book for what it was. Second, the book was published in Italian—and not Latin, like every other scientific work of the epoch—which showed that Galileo tried to reach as large an audience as possible. The pope felt personally betrayed for two reasons: first, because Galileo did not respect his wish for the two theories to be presented in a neutral manner, and second, because Galileo hadn't included a single fact to validate the Copernican theories he had defended so passionately.

The pope had to act, and fast—the book was quite a success. Although he wanted to save Galileo from appearing in front of the judges, the Inquisition commissioner refused. So Galileo was called to Rome by the Holy Office of the Inquisition in October 1632. Due to health reasons, he couldn't get there until February 1633. His theory itself wasn't why he was in trouble. First and foremost, he had disobeyed the will of the pope, and second, he had supported a banned theory. He was questioned. The pope even threatened to have him tortured if he didn't cooperate, and on June 22, 1633, Galileo was condemned.

The sentence:

> In Florence, there is a book called *Dialogue of Galileo Galilei Concerning the Two Chief World Systems, of Ptolemy and Copernicus* in which you defend the opinion of Copernicus. For your sentence, we declare that you, Galileo, are hereby made strongly suspect of heresy for having held this false doctrine that the earth moves and the sun does not. Consequently, with a sincere heart, you must renounce and curse before us these errors and heresies against the Church. And so that your great error does not go unpunished, we command that this *Dialogue* be forbidden by public edict, and that you be imprisoned in the prisons of the Holy Office.

And then the famous confession, prepared by the Holy Office:

> I, Galileo, son of the late Vincenzo Galilei de Florence, age seventy, being brought here for judgment, on my knees before the very Most Eminent and Most Reverend Cardinal General Inquisitors against all heresy in Christianity, having in front of my eyes and in my hands the Holy Gospel, swear that I have always held as true, and still hold as true, and will in the future, with the help of God, hold as true all that the Holy Catholic and Apostolic Church asserts, preaches, and teaches. However, since I have been enjoined by injunction of the

Holy Office to completely abandon the false belief that the sun is at the center of the universe and doesn't move, and that the earth is not the center of universe and moves, and to not defend nor teach this false doctrine in any way, oral or written; and after having been warned that this doctrine does not agree with what the holy scripture says, I wrote and published a book in which I wrote about this condemned doctrine and presented it with emphatic arguments, without refuting it in any manner; this is why I was held as highly suspect for heresy, to have professed and believed that the sun is fixed at the center of the universe and that the earth moves and is not the center of the universe. *With a sincere heart and true faith, I renounce and curse my errors.*

The pope immediately commuted his sentence to house arrest and Galileo was never put in prison. At first, he resided with Archbishop Piccolomini in Siena. After that, Galileo was permitted to return to his home in Florence, where he spent the rest of his life in blindness. However, a few months before he had gone completely blind, he wrote *Dialogues Concerning Two New Sciences*, which would be his last book. In this book, he laid the foundations of modern science, mechanics, and therein buried Aristotelian physics. Galileo died on January 8, 1642.

And Yet, It Moves . . .

After his retraction, some believe Galileo uttered, "Eppur, si muove,"[108] in one final burst of rebellion. This probably isn't true. In fact, if Galileo had truly said these words, he would most likely have been burned at the stake. The only other possibility is that he said these words at home, muttering them into his beard, far from the eyes—and ears—of the Inquisition.

---

[108] "And yet, it moves."

Galileo would be posthumously cleared, partially at first, then completely. In 1728, James Bradley proved that the earth revolves around the sun. After that, Pope Benedict XIV gave imprimatur for the first edition of the complete works of Galileo, stating however—by adding to the text—that the motion of the earth was only alleged. The convictions of 1616 and 1633 were not overturned. In 1757, Galileo's books were finally removed from the Church's list of prohibited books.

In the late twentieth century, Pope John Paul II didn't overturn the convictions, because that particular tribunal doesn't exist anymore. However, he did emphasize Galileo's genius, and in 1992 the pope said:

> Paradoxically, Galileo, a sincere believer, showed himself to be more perceptive in this regard than the theologians who opposed him. "If Scripture cannot err," he wrote to Benedetto Castelli, "certain of its interpreters and commentators can and do so in many ways."[109]

Better late than never, as they say, but sincerely, folks, sooner is better.

In the grand scheme of things, we are finally done with Aristotle's traditional philosophy of nature. Giordano Bruno has shaken up the generally accepted ideas about the universe—but without really proving anything. You might conclude from his works that, yeah, he had some good insights. But he'll always be the true father of relativity to me—and Galileo has essentially laid the foundations for mechanics.

What we lack at this point is someone with a mind that is as sharp as it is strong, able to see what others couldn't see, with unprecedented ability to prove the abstract, capable of revolutionizing the way we understand the world and the universe. Incredibly, this great mind will achieve a multitude of astounding accomplishments, including surviving an outbreak of the plague, only to be eternally associated with [a common fruit] . . . the apple.[110]

---

[109]   From *L'Osservatore Romano*, November 4, 1992. The November 21, 1613, letter is quoted from *Edizione nazionale delle Opere di Galileo Galilei* (1968), vol. 5, p. 282.

[110]   Sigh. Hang head.

50. Good Old Science Dude: Isaac Newton

To say Isaac Newton was a genius is just lazy writing. There is so much to say about him that this book couldn't possibly hope to cover the entire scope of his extraordinary knowledge, perseverance, and work. So instead, we'll limit ourselves to Newton's work in mechanics, but just so you know, he was also an expert in alchemy, theology, and exegesis[111] of the Bible . . . the list goes on and on! Even though I'll be focusing on his work in mechanics, I'll tell you right now, by the end of this chapter, I still won't have covered it all.

Isaac Archibald[112] Newton was born just about one year after Galileo died. He was born in January 1643—or on December 25, 1642, depending on the Julian calendar being used at the time, Merry Christmas! Joyeux Noël! His childhood wasn't a particularly happy one, however. Newton's father died a few weeks before he was born, and his mother remarried when he was three years old. He was then sent to live with his grandmother, and he attended the local school. When he was sixteen, his mother called him back to her so he could become a farmer and manage her land. Thank goodness, she noticed he was much better at mechanics than farming, so she agreed to let him go to school, and later, what the heck, college, too.

When he was eighteen years old, he started at Trinity College,[113] Cambridge, England. He studied mathematics in a big way—geometry, arithmetic, etc.—and he was really into astronomy. In 1665, when he was twenty-two years old, there was an outbreak of the plague in the city, forcing the college to close. It took two years for the plague epidemic to end—and the college remained closed for the duration. Newton spent those two years back at the farm. They were the two most fertile years for Newton's incredible mind. He made great strides in mathematics, physics, and optics. It was during this period that he first understood—and demonstrated—that white light wasn't actually white,

---

[111] Pronounced "ek-si-jee-sis": "explaining or interpreting."

[112] His middle name was absolutely not *Archibald*; I just like the sound of it.

[113] Of course.

it was all colors of light superimposed. In that rustic setting, he would also invent the beginnings of modern mathematical analysis, the branch of mathematics that grew out of infinitesimal calculus. We now call infinitesimal calculus simply "calculus," and it deals with limits, continuity, differentiation, and integration of functions.

In 1669, he finally returned to Trinity College. At this time, he replaced his professor as the Lucasian[114] Chair of Mathematics. In 1672—he was only twenty-nine years old—he was admitted to the Royal Society of London.[115] He also successfully built a telescope with a spherical mirror that didn't produce any chromatic aberrations—it didn't mess with the colors—which was unheard of! The next year, when he published his work on light, he became suddenly famous and the center of controversy.

You have to understand, Isaac Newton was not a great communicator. He spent much of his time completely alone—he almost always worked alone. He worked day and night, sometimes not even stopping to eat. He didn't have any social life or social network—he remained single his entire life. He was very reluctant to publish his work. Even when he did write a book, he wouldn't permit it to be published for years. That's what happened with *Opticks*, his work on the physics of optics. It was written in 1675, but it wasn't published until 1704, nearly thirty years later. In *Opticks*, he described how he used a prism to demonstrate the composition of white light. It's also in this work that he presented his corpuscular—particle—theory of light (see page 42).

In his quiet corner of the world, Newton worked on falling bodies, and an idea that came to him as he was sitting quietly in his mother's garden right next to the orchard, when he was observing the moon. An extraordinary event happened: A ripe apple fell from an apple tree!

---

114   Named for Henry Lucas, who made a donation to the college to finance a professorship for applied math. This same position was held by Stephen Hawking for thirty years in the twentieth century.

115   The British version of the French Royal Academy of Science. Fictionally, some of its members were seen betting against the character Phileas Fogg in the movie *Around the World in 80 Days* (2004).

Although you may have heard differently in textbooks and cartoons, the apple did not hit Newton on the head. No. The apple simply fell to the ground. And a question went through his mind: An apple, when released from the branch, falls to the ground. But the moon isn't held up by anything, so why doesn't it fall as well? That's the end of this little story out of nowhere, but in the next few paragraphs this question will be a very big deal.

In 1684, Edmund Halley—yes, he of the comet—asked Newton what he thought about Kepler's model of elliptical planetary motion. Newton shared his thoughts and where he was with his work on the subject. Hmmm, how can I explain how excited Halley was about what he heard? Ah yes, I know. He was so excited that he positively badgered Newton to publish the work, *plus* Halley covered the cost of publication. There you go. Therefore, in 1687, Newton published his "little" book, *Philosophiae Naturalis Principia Mathematica*,[116] which revolutionized the sciences for the next two centuries. You may have noticed I usually give the title of a book in English, then provide the original title in a footnote. But this book is so significant that it's known worldwide by its Latin name—often referred to as simply *Principia*.

We'll go through a few of the details—but really, just a few—about the set of theories he developed in this major scientific work. First, as the title indicates, the primary goal of the book is to put physics into mathematical terms—*natural philosophy* is the old-fashioned name for "physical science." Newton presented his theories and his principles, but as equations. *Principia* included what we now call the *laws of motion*—even though, actually, they're principles.

Newton's first law of motion, also called the *principle of inertia*, is as follows:

> Every object perseveres in a state of rest or uniform movement in a straight line in the same direction, unless some force acts upon it and compels it to change its state.

[116]   *Mathematical Principles of Natural Philosophy.*

What this law says is a generalization of what Galileo observed in his boat experiment—namely, there is no difference between a body at rest and a body in inertial motion—meaning moving in a straight line at a constant speed—and furthermore, what this law says . . . what makes it truly the principle of inertia . . . is that when a body is at rest—or in uniform rectilinear motion—its condition won't change unless a force is applied. In other words, if an object is sitting still and you leave it alone, it won't spontaneously move, and similarly, if a body is drifting along in space, far from the gravitational influence of another body, without any air to slow it down, it will continue on its course; it won't slow down and its trajectory won't change. This principle is basically the *fundamental principal of statics*, which states that if forces are applied to such a body, but they cancel each other out, everything continues as if no forces were applied at all. This first law of motion also allows us to define a Galilean reference frame as being a reference frame in which this law is valid.

Reference Frame, Galilean or Not

A reference frame is nothing more than what we commonly call *point of view*. In Galileo's boat experiment, an observer on the riverbank had a reference frame in which the boat was in uniform rectilinear motion as it went down the river. And Galileo, as an observer in the cabin, had a reference frame in which the boat was at rest, because it did not move *relative* to the reference frame, meaning *relative to Galileo*.

Some reference frames aren't Galilean. For example, if the boat had accelerated, Galileo would still have a reference frame, but in this one, a ball sitting on the floor would start to roll, even though no force had been applied to it. In this reference frame, Newton's first law isn't valid, so it's not a Galilean reference frame.

Newton's second law of motion, also occasionally called *the fundamental principle of dynamics*, is as follows:

Changes in motion are proportional to the driving force and are in a straight line, in the same direction as the force was applied.

First and foremost, this law introduces the concept of *acceleration*. I'm sure you know what acceleration is. I'm also sure it may be helpful to clarify what acceleration means in physics. Acceleration is change in velocity over time; this includes deceleration too. In physics, deceleration is acceleration with a negative value. Velocity is the change in the position of a body over time. Since you know speed is calculated in meters per second or miles per hour, this clearly shows that a speed tells you the number of meters traveled in one second or the number of miles traveled in one hour.

Newton's second law says: A body to which a set of forces is applied accelerates in proportion to the resultant of these forces. To say it more simply, if you apply one force pushing forward and another force pushing left on a body—let's say a ball—the ball will experience acceleration to the front and to the left. It certainly seems as though I'm making a mountain out of a molehill here—and apparently, I seem to be really enjoying making this molehill into a mountain for no particular reason at all—but, folks, the fact is that this law made it possible for Newton to explain how the moon continues to revolve around the earth instead of falling and crashing into the earth. More details soon. I promise.

You'll notice that Newton's first two laws provide the *fundamental principle of statics* and the *fundamental principle of dynamics*. So, really, I wasn't screwing around when I said that this book revolutionized physical science.

Newton's third law of motion, sometimes called the *law of reciprocal actions*, is as follows:

The action is always equal to the reaction, meaning that the actions of two bodies on each other are always equal and in opposite directions.

This law says that when you apply a force to a body—the action—this force also applies a force of equal magnitude and in the opposite direction on you—the reaction. It's the fundamental principle for how we get our spacecraft to move in space. In space, if you want to go in a particular direction, you project gas in the opposite direction, because when you apply a force to push the gas out, the gas applies an equal force on your spaceship, in the opposite direction, and you move—it's the principle that Wall-E[117] the robot and, more recently, Sandra Bullock[118] used to propel themselves using a fire extinguisher.

Together, these three laws are sufficient for explaining a crapload of things about bodies in motion, except for the most common [situation], falling bodies. Newton is about to achieve immortality.

Newton's law of universal gravitation is as follows:

> Two bodies with mass attract each other in a way that is directly proportional to the product of their masses and indirectly proportional to the square of the distance between them.

This law says two bodies that have mass attract each other. Period. Wherever they come from and wherever they are. As you read these lines, your very own body is attracting the entire galaxy of Andromeda. And that's not nothing. Well, the attraction *is* extremely small when you calculate it, but it's not *nothing*. All bodies that have mass attract each other. And this law tells us that the intensity of the attraction is proportional to the masses—if you are twice as heavy, you attract twice as much—and inversely proportional to the distance squared—if you move apart, the intensity of the attraction decreases. If you double the distance, you divide the magnitude by four; triple it, and you divide by nine. Obviously, the attraction between two bodies diminishes rapidly with distance.

Why doesn't the moon fall? After all, Newton had just proved that there's a force attracting the moon toward the earth—equal to the force attracting the earth toward the moon—so why doesn't the moon fall? To answer this question, Newton used the analogy of a cannon. If you fire a cannonball, say to the east, the cannonball will travel a certain

---

[117] In the movie *Wall-E*, Pixar Animation Studios (2008).
[118] In the movie *Gravity*, Warner Bros. (2013).

And Where Is Hooke in All of This?

There was a long-standing dispute between Newton and Hooke, the kind they go crazy about on the internet, so eager are they for stories of secrets and conspiracies; there would later be a similar situation concerning Einstein and Henri Poincaré. What was the dispute about? Some say Robert Hooke, secretary of the British Royal Society, presented the universal law of gravitation before Newton did and got absolutely no credit for it.

But before we go any further, you need to understand that these two guys did *not* like each other. They were constantly arguing about gravity and light. In fact, Newton waited until Hooke was dead to publish *Opticks*. That's definitely one way to have the last word. Besides just arguing about theories, Hooke accused Newton of working on gravity on the sly while Hooke himself was working on the subject. Hooke was out of his mind with rage. He accused Newton of stealing the inverse square theory. Newton categorically denied any knowledge of Hooke's work. Today, we know Newton lied about having seen Hooke's work, but it wasn't to hide guilt; it was just because he really hated the guy. In 1674, Hooke had, in fact, correctly formulated a law of attraction that was similar to the one Newton presented ten years later. But you see, Hooke hadn't provided any proof and was never able to validate his inverse square hypothesis. It wasn't possible to validate it until Huygens published his laws of centrifugal force. As for Newton, he found the inverse square law by applying Kepler's third law—as he had shown Halley in 1684.

Be that as it may, we should still give Newton credit for wanting to validate his early mathematical theories—as early as 1666—by experiment, to confirm the universal nature of gravitation, by measuring the attraction of the earth on the moon. Since he couldn't get a conclusive result, he put his theory aside until 1682. That's when he heard that Jean Picard of France had calculated a more accurate radius of the earth. With this better value, he got results that matched his theory. And that validated his theory of universal gravitation!

distance to the east before crashing to the ground. If you fire with more power, it goes farther. If you fire higher with more power, it goes even farther. And if you fire so high, and with so much power, that when the cannonball starts to fall back down, the curvature of the earth falls away below it, then what? It goes even farther until the cannonball finally lands back on the ground. But with enough speed and enough height, the cannonball continues to go around the earth and never hits the ground. It would keep going straight off into space, except the gravity of the earth pulls it back toward Earth. The cannonball keeps falling and missing, and the earth keeps pulling it back in line. Thus it stays in orbit.

That's exactly what the moon does around the earth. It's always falling, which accelerates its motion. At the same time, it's moving away from the ground because of its horizontal speed. As a result, the moon continues to orbit around the earth. And that's exactly what the earth does around the sun, and what all the other planets do around the sun, and what all moons do around their planet. That's how celestial bodies move: They are all continually falling somewhere.

And of course, Newton's laws matched Kepler's empirical data, which validated the laws very nicely; and people—the species, not the magazine—considered Newton's work as the crowning glory of scientific endeavor. No one questioned Newton's work for more than two hundred years. It would take a Maxwell and an Einstein to shake the blind faith in Newton's laws. So it's really no surprise that classical mechanics is still commonly called *Newtonian mechanics*.

51. Force, Couple, Torque, and Work

> The Force gives a Jedi his power. It's an energy field created by all living things. It surrounds us and penetrates us. It binds the Galaxy together.[119]

Well, *a force* is nothing like that. There, I said it. You can just get the *Star Wars* idea—as cool as it is—out of your head, because that's not what we're going to talk about here. In the next few chapters, we're going to be talking

[119]   Obi-wan in *Star Wars: A New Hope* (1977).

about velocities, accelerations, and motion as well as influence—effects, reactions, and interactions, and more. But wait, I just noticed, we haven't even taken a few lines to define these things. That's because they are so obvious, right? Not exactly. Though we might think we have a generally good idea of what a force is, we tend to make the definition way too broad, and then it covers other things as well. So let's read the next few lines.

In mechanics, a force is an interaction between multiple bodies. A body exerts a force on another body when it acts on it, whether by deforming it, attracting it, pushing it, or something else. We typically show a force as a vector—it's a mathematical item that's very useful. It seems to have been invented specifically for this purpose—at least, that's nearly the case. In *Principia*, Newton set up the basis for vector analysis. A force vector is defined by four characteristics: point of application, direction,[120] line of action, and magnitude. Showing it as an arrow is perfect for a model, but it doesn't match reality.

For example, when you hit a ball with a bat, there isn't any "point" of application in the mathematical sense of an "infinitely small point." The contact between the ball and the bat is actually distributed over an area—and when the ball comes in contact with the bat, the ball *and* the bat both deform. But modeling in mechanics has the elegance to use relatively simple mathematical tools to describe interactions that are, in reality, extremely complicated.

However, you'll notice right away that there are big differences between rigid solid body mechanics—well, considered rigid anyway—and flexible body mechanics. I'm sure you can understand there's definitely a difference between hitting a bowling ball and hitting a towel. And then there's fluid mechanics—mechanics of liquids and gases; naturally, they behave according to their own rules, but they still obey Newton's laws.

When multiple forces come into play, we can just add them together to calculate what is called the *resultant force*—if you draw the forces as

---

[120]   The line of action is a line—let's say, vertical—of indefinite length, and the direction (positive or negative) is the way the force points along this line—let's say, downward.

arrows, all you have to do is put them all end to end to figure out the resultant force; as I said, these vector things are very useful tools.

And most of the time, when a body is acted on by a force, let's say to the right, it causes motion to the right. A force pointing to the left? Motion to the left, and so on. I say "most of the time," because sometimes a constraint prevents a body from moving. Imagine a revolving door in front of you. You push on the door by applying a forward force. But the door doesn't move forward; it turns. Of course, we can explain this with what we already know. We know that as the door moves forward, the center shaft keeps the whole thing from moving forward, so only the moveable side moves forward, and each instant, the force will make the movable side of the door move forward at a slightly different angle, thereby converting forward motion into rotational motion. And this conversion of translation (straight line movement) into rotation leads us to two new concepts: couple and torque.

A *couple*, as its name indicates, is made of two forces—and we can extend the idea to cover more than two forces, but it's always called a couple—with equal magnitude, but opposite directions, applied to the same body but not applied at the same point. Considering it one way, the resultant—the sum—of the two forces is zero—that's all we expect from a couple . . . well, that and a rotational motion (note that there's no translation in the motion). If you have a cylinder in front of you that can turn on a vertical axle—like those prayer wheels you see in Tibet and Nepal, or like those spinney tic-tac-toe games on the playground—you can turn the cylinder by using both hands. Your left hand pushes the left side away from you, and your right hand pulls the right side back toward you. The resultant of the two forces applied by your hands is zero, yet the

Engine Torque

In a car, the engine torque is the rotational or twisting force applied to the crankshaft in the engine; the force is applied *by* the piston rods (connected to the pistons) that push on alternate sides of the shaft—like pushing on the pedals of a bike—as it spins. The larger the torque, the faster the engine can produce power.

cylinder acquires a rotational motion. How can two forces that cancel each other out cause motion? It's because these forces are applied around an axis of rotation. This produces a *torque*.

The *moment* of a force relative to a point—the moment of a force is always relative to a point—reflects the ability of the force to make a body rotate about that point. The magnitude of the moment is proportional to the force's distance from that point—that distance is called the moment arm, just FYI. OK, let's get up and stand in front of an open door—just any regular ol' door that swings on its hinges. Close the door by pushing on the edge that's farthest from the hinges. The door closes easily, right? Now, try to close the door—OK, smart aleck, first, open the door again. Now try to close the door by pushing on the edge right next to the hinges—its axis of rotation. It's a lot harder, isn't it? So you've just demonstrated that, as the force is applied farther and farther from the door's hinges, the moment relative to that axis of rotation increases. Although he didn't say it quite this way, Archimedes understood the concept when he explained the principle of the lever: the longer the lever arm—that is, the greater the distance between the force and the pivot point[121]—the more effective your efforts will be.

The *work* of a force is the energy provided by the force during displacement (motion). When the force is in the same direction as the displacement, we call it positive work—like when you push a car up a hill. When the force is in the opposite direction of the displacement, we call it negative work—like when you walk a bike down a hill—you're holding it back. Finally, if the force is applied perpendicular to the direction of the movement, so it's not even partially in line with the direction of the movement in either direction, we talk about zero work—for example, if you push down on an electric kiddie car while it's moving across flat ground. Now, you don't want to confuse zero work with a force that has no effect. Think about this: As the moon travels around the earth, it is always moving tangentially (trying to go off along a

---

[121]   Another word for *pivot point* is *fulcrum*; say it . . . "full-crumb." It just feels so good in your mouth and sounds so powerful when you say, Give me a fulcrum and I will move the world!

tangent line into space) because of inertia.[122] The moon is also constantly subject to the force of gravity, which is applied vertically from the center of the moon toward the center of the earth, so gravity's force always acts perpendicular to the motion of the moon. So this force—gravity—does zero work! But it certainly has an effect. Without it, the moon would no longer orbit the earth!

52. Momentum and Collisions

We're about to get into some pain-in-the-butt definitions in mechanics, but it's important to take the time to understand them. Once you grasp the concepts, you'll have a lot better idea of not only classical mechanics but also quantum mechanics—and believe me, when it comes to quantum mechanics . . . anything helpful is a good idea.

What we call *momentum* is defined as the quantity of motion of a moving body. It's measured by multiplying the body's mass times its velocity. OK, to be a little more rigorous than that, we'll have to dig a little deeper into the concept. Let's use an example that is the most in, the hottest, the most incredibly hip of any possible example: shuffleboard. You shoot your *cue* disc at another disc—you might be trying to make a *bunt*, moving your disc to a more favorable spot, but what you really want is a *bunny*, the disc in the winning location—anyway, you shoot your shuffleboard disc and it makes a direct hit on a *target* disc. When the cue disc hits the target disc, which is at rest, the cue disc stops abruptly, and the target disc instantaneously starts moving and appears to continue the motion of the first disc. We certainly understand that something was transmitted from the cue disc to the target disc. What can we say about this "something"? We can say it depends on the speed of the moving object. We can well imagine that if you had launched the cue disc three times faster or considerably slower, the target disc would have taken off faster or slower, depending on how fast you shot that cue disc. We can also say that this "something" depends on the mass of the disc that was launched: If you shot a manhole cover, or, say, an

---

[122]   Newton's first law in use here; see page 181 for a reminder.

air-hockey puck at the target disc, you'd certainly see that the target disc would take off all the faster and harder when it's hit by a heavier disc.

This "something" is called *momentum*. It quite exactly equals the mass of a body multiplied by its velocity. Velocity is described by a vector that has a magnitude and a direction; thus momentum is also a vector. Frequently, momentum is confused with the notion of impulse. But impulse was actually one of those early theories on the way to understanding momentum. Look at Newton's cradle—you know, that thing with five steel balls hanging next to each other that you usually see on an executive's desk in the movies and that makes that infernal "tic, tic, tic" sound—if you pull the left ball out and let it go, as soon as it hits the next ball, it comes to a complete stop. And the ball all the way on the right side continues the motion. Then it swings back and transmits its motion to the ball all the way to the left. In this case, momentum is transmitted from ball to ball—to ball to ball to ball to ball to ball . . . I hate that noise. Momentum is a particularly handy property to use in the study of collisions. Because when you have a collision, there's a transfer of momentum. But that's not all.

In physics, the study of collisions is an important branch of mechanics. And there are various types of collisions—a collision is defined as an impact between two bodies. There are *elastic collisions*, *inelastic collisions*, and *partially elastic collisions*, and some are completely "not elastic"; they have a really elegant name for that kind of collision: *perfectly inelastic collisions*.

So far, we've been talking about ideal, aka perfect, situations in which a body is represented as a point, the surrounding air had no effect on the motion, and there's no friction from the floor, etc. Obviously, reality is not like that. Fortunately, in a good number of the cases, we can consider some effects as negligible—like air resistance on a small falling ball—and it seems like an ideal condition, but remember in the back of your mind . . . that's never the case. For example, an elastic collision doesn't exist. Ever. Except maybe, on an atomic scale, but that's a special case, which is definitely beyond the framework of classical mechanics.

An elastic collision is a collision during which all of the momentum is transferred from one body to another without any loss of momentum, and no energy is lost during the impact. First of all, for an impact to be completely elastic, we must assume the bodies act in an isolated system, because in an isolated system, there is conservation of momentum.

Isolated System

An isolated system is a physical system—such as a group of balls—that doesn't interact *at all* with its environment. In reality, an isolated system can't exist, because all mass in the universe exerts a gravitational force, though perhaps minuscule, on every other massive body in the universe. Thus we shouldn't automatically think of the earth as an isolated system, because it's definitely influenced by the sun. And the solar system can't be considered an isolated system due to the influence of the stars in our galaxy. And our galaxy can't be because of the influence of the other galaxies. It may be possible—it remains to be proven—that our universe *as a whole* is actually an isolated system. If such is the case, that makes it the only universe. All that said, it is common practice in physics to simplify situations in order to model them. We have to think about it first, then decide if considering something as an isolated system is a reasonable thing to do. And often, it is. Thus we can decide to consider a ball rolling on a slanted board as an isolated system, consisting of the ball and the board.

Next, we have to imagine that during an impact, all energy is transmitted only as kinetic energy—the energy from being in motion. In reality, that's never how it happens. An impact, in reality, generates heat. The heat energy comes from the kinetic energy of the bodies before impact. Usually, some energy gets converted into heat—we say it dissipates: It dissipates as heat—at least in part. However, in a completely elastic impact, no kinetic energy is dissipated. If two bodies with the same mass moving toward each other have a head-on elastic impact—heading straight toward each other and hitting dead center—each one bounces back in the direction it came from—and this is so entertaining: Each one goes off at the speed the *other* body was traveling before impact. That last little entertaining part was due to the fact that the two bodies had the same mass. Because remember, momentum depends on mass.

In the same way, again with two bodies of the same mass, if one is sitting still at the time of impact, it stops the motion of the other body, and it takes off in the same direction and at same speed that the other body was traveling before impact. It's what you observe over short distances—before the effects of friction come into play—on a pool table.

Inelastic and partially elastic collisions match reality more closely. When two cars hit each other, part of the energy dissipates as heat and part as noise—yes, it takes energy to produce noise—and some energy is also used to bend the fenders on impact. Then the rest of the energy is transferred as kinetic energy, making the cars move according to their mass, speed, and angle and the intensity of the impact. This case may sound closer to reality, but it's very complicated to model, if only due to the possible deformation of the bodies. You can't imagine the amount of time and effort engineers—bearded and bespectacled, or not, but always with a pen in their pocket protectors—put into the design of a bumper or a shock absorber.

Take a basketball—I mean mentally, don't go looking for a basketball . . . or actually, go ahead, why not? Go get one and make two or three historic baskets. Sports are good for you—OK, now, take a basketball, and hold it at some height. Put one hand on top and drop the ball. Don't move your top hand when you drop the ball. Notice the ball never bounces high enough to touch your hand. In fact, each subsequent bounce has less height. That's because the collision between the ball and the ground is inelastic. Every time they make contact, the ball deforms at little bit, this compresses the air inside, and the compression heats the air up a little bit. The increased temperature increases the air pressure, which pushes on the bladder of the ball, which helps it—at least somewhat—to bounce better.

And last, but not least: perfectly inelastic collisions. Here, we have the opposite of perfectly elastic collisions. In this case, the maximum amount of energy that can be lost by dissipation is actually lost. Also during a perfectly inelastic collision, the bodies stay together after impact, aka a soft collision. Because it's an isolated system, the total momentum is conserved; however, all energy that can be dissipated—as heat, sound, or deformation—is effectively lost. After impact, the system's kinetic energy is lower. Frequently, in a two-car, head-on collision, the cars are stuck to each other after impact—they collide, then slide down the street together—that's an example of a perfectly inelastic collision. Now consider the case of two bodies with the same mass, m. One body is moving toward the other at an initial velocity, v, the other body is not moving (aka at rest). They have a perfectly inelastic collision, and together, they follow the same trajectory as the body that was moving, but at half the initial velocity.

After all that, we still want to talk about angular momentum and its conservation, because it happens to be why the solar system is flat *and* how stunt car drivers catch awesome air and still land flat on all wheels—true—and then we'll generally have covered classical mechanics. OK, well, not fluids[123] and not thermodynamics, but they are really completely separate subjects.

53. Angular Momentum

OK, I tell you no lies—angular momentum is hot! Mathematically, it's a very powerful tool—a giant wrench—but seriously, the reality of angular momentum is rather twisted. So, with apologies to the mathematicians and physicists who might get some gray hair from reading the next few lines, I'm going to try to simplify the problem.

When a body moves in a straight line, it has momentum. Angular momentum is, sort of, a form of momentum but in rotation. At a minimum, angular momentum "behaves" in rotation the way momentum "behaves" in translation (moving in a straight line). Specifically, in an isolated system, angular momentum is conserved—just like regular ol' linear momentum is conserved. Just to clarify a little, we talk about the

[123]   Ha, ha! *Well* . . . fluids . . . *well*—fluid mechanics—well, I thought it was funny.

angular momentum of a point—on a rotating body—relative to another point—the axis of rotation, or the pivot point.

Without getting too deeply into what angular momentum is, knowing that it's conserved gives us a better feel for what it does, especially as it pertains to our solar system being flat. As you already heard (see page 100), when the solar system was beginning to form, there was a nebulous—or more accurately, a molecular—cloud. It contained a vast number of particles . . . gas particles, ice particles, dust particles. All these particles were moving here, there, and everywhere, colliding with each other, spinning and revolving around each other. Even with all this random motion, if modeled particle by particle, the model would show that the cloud also moved as a whole. And as a whole, it was rotating. If we consider the molecular cloud as an isolated system—which is a reasonable thing to do—we'd realize this rotating motion produces angular momentum that will be conserved as long as the system remains isolated—by the way, it is still conserved to this day, 4.5 billion years later. When the particles started to clump together due to gravity, the middle-*ish* section—the part that was getting bigger—emerged and became our sun. It spun faster and faster, bringing the rest of the cloud along for the ride. The cloud began to draw out and flatten, into a large disk—this allowed for the conservation of the angular momentum of the system. That's why we can observe, some 4.5 billion years later, that our solar system is—more or less—flat.

A striking, more easily observable, and to some, far more exciting example is when a stunt car driver uses conservation of angular momentum. I'm sure you've seen a movie where a car hits a rise in the road—or maybe a poorly concealed ramp under the frame—and it catches some air, literally flies through the air, sometimes for several yards . . . actually, it's always several yards. The laws of physics tell us that when a car leaves the ground with the front of the car angled toward the sky, it is expected, in the best-case scenario, to land with the front still angled toward the sky. Or in a not-so-best-case scenario, it might increase its angle with respect to the ground and land vertically. Or in a worst-case scenario, it could land completely upside down!! But in the movies, because it's more photogenic and more practical, we frequently see the car landing flat, or maybe with the front of the car tilting slightly down-ward. How does that work?! Conservation of angular momentum makes this feat possible. When a car leaves the ground, it becomes an isolated system—specifically, isolated from the ground. Its wheels are spinning, and this rotation produces angular momentum. Once in the air, the stunt driver hits the brakes to lock the wheels. At that point, they are no longer producing angular momentum. But angular momentum must be conserved. So what happens? When the brakes stop the wheels from spinning, the whole car starts to rotate—more slowly than the wheels did, because the whole car is heavier, so it has a lot more inertia. This conserves the angular momentum. Thus the laws of physics are satis-fied, and Michael Bay is one or two explosions from being completely satisfied with what he sees. That's a wrap!

And that's all, folks! At this point, you have had the grand tour of classical mechanics. Along the way, I've tried my best to honor Gior-dano Bruno, Galileo, and Isaac Newton. Some might say I should have spent more time on Nicolaus Copernicus, if only to mention that he prudently waited until he was on his deathbed to publish his work on the heliocentric solar system, thus avoiding any nonsense with the Church. I could also have emphasized Kepler a bit more. I'm sure there are some who are appalled that I barely even mentioned Descartes and Leibniz. And of course, others will complain that I didn't cite Euclid, or Pythagoras, or Thales, or Euler—let's face it, I'd have to write a com-pletely separate book on mathematics to give you even a glimpse of the extent of their work. I definitely could have gone a little farther afield, leaving Europe behind to talk about sciences developed in China, India, and the Middle East. So many perspectives, so many stories to tell; some are moving, some thrilling. Others show, through the bitter and some-times repeated failures of the protagonists, that a failure is not the end, it's actually a step toward progress. Winston Churchill himself said that success is going from failure to failure without losing your enthusiasm, and Churchill certainly knew a thing or two about success.

Obviously, I had to make some tough decisions to skip over some subjects, at least for the time being, so we could move forward in our journey. Now that classical mechanics is behind us, the pieces are in place for us to talk about relativity. But before we get there, as an interlude, let's spend a little time talking about another, possibly even more fascinating, kind of mechanics: that of the human being. And afterward, I promise, we'll get to Einstein's theories.

LIFE

We've never been just beta version 1000000001rc.

Now, before we try to understand what makes us *thinking* beings, let's ask ourselves what makes us *living* beings? I am alive—at least, at the moment, I'm alive. I don't know what condition I'll be in when you read the book . . . ouch—but you're alive right now. I know it and you know it. But what does it mean to be alive? Can you prove—and if so, how?— that everyone around you: your family, your neighbors, your hairstylist, the guy at the deli, me, Bernie Sanders—aren't all just figments of your imagination?[124] OK then, let's accept—frankly, that's all we can do—that they and we are alive; so how do we know we are alive? How about your dog, the spider hanging out on the ceiling, discreetly eliminating the mosquitoes that would otherwise be taking a daily blood sample from you, the aforementioned mosquitoes, the earthworm I once impaled on a hook, the fish that ate the worm . . . all these animals . . . do they know that they are alive? Ultimately, is life nothing more than . . . this? Can it be any other way? And regardless of the answer to that last question, can life exist elsewhere?

Welcome to the part of the book that asks a gigantic amount of questions, especially considering how few answers we get.

---

[124] If you are Bernie Sanders, it goes without saying, when you read this you'll have to pick some other celebrity as an example and substitute that celebrity's name into the question.

54. You Are Alive

You know it, no need to go over that again. But what makes you alive? You might reply that you have a conscience, that you have a pulse, that your blood circulates in your veins, you breathe, you eat, you digest, you grow, you can reproduce, you were born, and you will die. All these things may be true, but are they enough to define what is alive? Are those the criteria? For example, a sterile human being can't reproduce but is no less alive. And even trees breathe—without lungs. They don't have a pumping heart, but they are still alive. They are indeed living things. Does a mosquito have a conscience?

The reality is that the definition of a living thing is primarily empirical—I've used this word before, it means based on actual experience—for most of us. And we are fairly able to identify what is alive—the dog, the cat, the goldfish, old Aunt Gertie who rambles on at Thanksgiving dinner—and what is not—the fork, the chair, the light bulb in the desk lamp, the moldy old stuffed moose head that belongs to old Aunt Gertie, who rambles on at family reunions. It's an ancient question, of great interest to ancient Greek philosophers, of course. But it has also fascinated scholars from every age and region, from Claude Bernard to Erwin Schrödinger[125] and Kant and Descartes before them.

To get a better grasp, we could read through the various definitions offered over the centuries about what was to be called life, and then apply any new scientific discoveries that give us new and improved insights. These are updated on a nearly daily basis. But let me tell you, that kind of project would make for a whole other book: a synthesis of philosophy, theology, science, and intuition. To narrow things down a bit, I suggest for the time being we not ask any questions about *breath of life* or the *soul*—sorry, Aristotle, another time perhaps—and let's focus only on the scientific aspects of life.

Without getting into a bunch of academic details—and assuming it's even possible to come up with an accurate definition life—our definition of life is going to be based on the following statements:

---

[125]   Made famous by a particular story about a cat . . . it's a quantum mechanics thing.

A living thing starts by being born, in one way or another.

It has a complex internal arrangement to maintain itself, or possibly transform itself, in a controlled manner.

It nourishes itself and grows, possibly feeding off the matter around it.

It can reproduce.

It will eventually die.

Thermodynamic Equilibrium

Thermodynamic equilibrium is one of the fundamental concepts of thermodynamics, the branch of physics that's concerned with heat and thermal energy transfer. Thermodynamic equilibrium is a fundamental concept because it defines the final state of any isolated system. If you put ice cubes in hot soup, the ice cubes will melt and the hot soup will cool down. And after a sufficiently long[126] period of time, the melted ice and the soup will reach thermodynamic equilibrium, meaning that, overall, nothing else is going to happen, whether in terms of heat exchange—it will all be at the same temperature—or in mechanical terms—it will all be the same pressure—or in chemical terms—the concentrations remain the same. An isolated system—which is something you don't mess with—will always end up in equilibrium. That's one of the fundamental laws of nature itself.

Once it's dead—and by the way, we are not going to get into defining what death is—a living thing doesn't do all that living business anymore. So it seems we've managed to develop a relatively clear, though somewhat superficial, idea of what a living thing is.

One key characteristic of living things is that they seem to violate the second law of thermodynamics (see page 182). And that's of interest to us, because it's something that can be described with a mathematical model. Schrödinger—other than wanting to torture cats that were both dead and alive (a nod to quantum

---

[126] This "sufficiently long" part bothers me because it's vague, but it really is part of the definition.

Entropy

We'll be discussing more details about entropy later, but the key thing at this point, as we begin a tiny little side trip, is to *get a feel for* what it's about. Take a college student. We'll say that from this point on, his studio apartment is an isolated system.

Now, you don't do anything. You don't try to control anything, except the first day of the experiment, you have cleaned up the room. There isn't a speck of dust on the furniture, his clothes are folded and carefully put away, and any food is stocked neatly in the cabinets. Leave him and his studio apartment alone for a week—depending on the student, a few days may be sufficient— and then come and check things out.

Upon returning to the apartment, you find there's dust on every surface that isn't protected by socks—yes, and there are plenty of socks and as you might expect, none of them match. Is there a law of physics for that phenomenon: the law of unmatched socks? Hmmm, that might be part of entropy, too. . . . There is still food aplenty . . . on the floor . . . that is, as long as you are still OK with calling it food after it starts to move around on its own. And a strange odor permeates the air; you can't quite put your finger on it. . . . Even though you've never been to Africa, you're pretty sure it is the compelling fragrance of a dead zebra that's been out in the hot savanna sun for a few days.

That's sort of the idea of entropy. The idea is that when there is no mechanism in place to control the organization of a system, it can become only more disorganized. Bottom line: The first law of thermodynamics says, if you put an ice cube into a cup of burning hot coffee, you'll end up with a lukewarm café Americano, and the second law of thermodynamics says, the lukewarm café Americano will never spontaneously change into burning hot coffee and an ice cube.

physics)—wanted to develop a formal and scientific definition for life. It wasn't a trivial question. He wanted to know if life obeyed the same laws of physics as the rest of the universe. It might seem like a strange question to you, but it's not all that crazy. Even though it's blatantly obvious that a living thing on Earth is subject to Earth's gravitation, same as a rock, there are more subtle laws of physics that don't seem to apply to living things—namely, the laws of thermodynamics.

Living things move away from thermodynamic equilibrium, which led Schrödinger to say,

> living matter, while not eluding the "laws of physics" as established up to date, is likely to involve "other laws of physics" hitherto unknown, which, however, once they have been revealed, will form just as integral a part of this science as the former.[127]

Let's take a quick look at the second law of thermodynamics—that's the law about *entropy*. It says in an isolated system, entropy can only increase.

Schrödinger noticed that a living thing, as a closed system, did not obey this principle, even though it's a fundamental principal of nature. And though it's often called *Schrödinger's paradox*, it's a paradox that is easily resolved.

In fact, Schrödinger himself quickly concluded that the only way to resolve the problem and not violate any law of nature is to say that a living being cannot be considered as an isolated system. That might seem like no big deal, but it certainly underlines the notion that a living being cannot be separate from its environment.

Well, that's all very lovely, but our original "definition" of life mentioned respiration, blood, reproduction, eating, and digestion, etc. Is that all valid?

And if so, why should we care?

55. The Incredible Highways and Byways of the Body

Breathing is definitely one of the first things that come to mind when we start talking about being alive. I'm sure that's because in literature, theater, and movies, the dead can be recognized as dead by virtue of not breathing anymore. That's because breathing, aka respiration, is the

[127]   Erwin Schrödinger, *What Is Life*.

most widely used method in the vast family of living things to get oxygen from the air so we can produce energy. In humans—and other mammals and birds—respiration involves filling and emptying the lungs. This is called pulmonary ventilation. And as everyone knows, it can be broken down into two actions, inhaling and exhaling. During inhalation (sucking in of air), the movement of respiratory muscles, such as the diaphragm, causes an influx of air to fill the lungs with oxygen-rich air—the regular air around you is about 21 percent oxygen (O_2). What happens next? The lungs are made of a huge membrane folded back on itself with many convolutions. Lungs are shaped a lot like an upside-down tree with a large trunk, which separates into large branches, which separate into smaller branches, which separate into even smaller branches, until you finally get to the leaves—in a lung, these structures are the trachea, bronchi, bronchioles, and finally the alveoli.

It's no random coincidence that lungs are shaped like trees, because trees breathe through their leaves. As a matter of fact, the tree shape of the lungs with all those numerous folds is the best way to obtain the largest possible surface area within the limited space of a human rib cage. There is a branch of mathematics called *fractal theory* that endeavors to understand the theory behind containing an infinitely large surface within a finite volume. Well, as it happens, our lungs are not an infinite surface. The surface area of an average lung can range between 540 square feet and 800 square feet. That means one lung would cover an area the size of a racquetball court. When the lungs are full, all that surface area comes in contact with the air and the oxygen in the air—on one side—we'll call this side "outside" of the body—in the sense that there's no divider between the outside of the body and this "outer" side of the pulmonary membrane. And on the other side of this membrane, there are blood vessels—lots and lots of blood vessels.

Here is where a phenomenon occurs that's completely passive and very commonly happens naturally on its own: When a permeable membrane separates two areas, and there's a higher concentration of something—oxygen in this case—in one area, an exchange occurs, and the "something"—if small enough—will cross through the dividing membrane from the area with the higher concentration to the area with the lower concentration until the concentrations on both sides are equal. This phenomenon with the self-explanatory name of *diffusion* is completely passive and spontaneous. Our body doesn't have to *do* anything in particular. Because the oxygen concentration in the air is much higher than the oxygen concentration in blood *and* because O2 molecules are small enough to pass through the pulmonary membrane, they spontaneously cross the pulmonary membrane to join the blood.

Le Chatelier's Principle

Henry Le Chatelier was a French chemist when the nineteenth century rolled over into the twentieth. He owes his fame to a principle he defined, in 1884, on the basis of many observations. This eponymous principle, Le Chatelier's principle, aka the equilibrium law, says the following:

> When a physicochemical[128] system in equilibrium is exposed to external disturbances causing it to a change to a new state of equilibrium, the change works against the disturbance that caused it in order to moderate the effect.

In other words, when a change from outside a system causes an imbalance, the system will spontaneously try to return to its initial state of equilibrium. In the case of our breathing, when we bring oxygen into the alveoli, it creates an imbalance in the oxygen concentration between the two sides of the pulmonary membrane. The "air in the lungs–pulmonary membrane–blood" system spontaneously tries to get back to its initial equilibrium by letting oxygen cross the membrane.

Once the oxygen[129] is in the blood, then what does it do? To answer this question, we're going to need to discuss—just a little—the

---

[128]   Pronounced: "Fizzick O'Chemical"—like some nice Irish guy who works in a lab.

[129]   Yes, as you noticed, we rarely bother to make the distinction between elemental oxygen, O, and molecular oxygen, O2, aka dioxygen or diatomic oxygen gas. We just say oxygen and let everybody else figure it out. It's not correct, but that's how people talk.

composition of blood. As you know—if you've ever cut, injured, or slashed yourself—do not do that, it hurts—or if you happen to be a woman of childbearing age who has experienced the monthly joy of being reminded of one of your body's biological features—your blood is liquid. And this liquid is red. Your blood is liquid, because it's primarily made of plasma, which is primarily made of water—more than 90 percent water, in fact. When separated from blood, plasma is a pale yellowish color due to the other 10 percent: trace minerals, nutrients, metabolic waste, hormones, etc. To be specific, blood is approximately 55 percent plasma. So, plasma is blood's main ingredient, but not by much. The remaining 45 percent—where blood gets its red color—consists of formed elements of the blood, aka corpuscles or blood cells.[130] That's where we find red blood cells, white blood cells, and platelets. Right now, we're particularly interested in red blood cells, sometimes called hemocytes, or even erythrocytes.[131]

When oxygen passes into the blood, some of the oxygen—a negligible part—will just dissolve into the plasma, but the vast majority of it will stay right there patiently waiting at the opening—right there in the "doorway"—waiting for the next unsuspecting red blood cell to come along. Cue the scary music.

For now, the important thing is that a red blood cell doesn't have a nucleus[132] even though it did have one when it first came into existence. A red blood cell is the result of *erythropoiesis*, erythroblast differentiation. An erythroblast is a cell that has a nucleus. This is important to know, because a cell's DNA is in its nucleus. Having lost its nucleus, a red blood cell doesn't have DNA anymore, and without DNA, it can't divide to reproduce itself. A red blood cell has only what normally surrounds the nucleus, the cytoplasm, which—in this case—is rich in hemoglobin.

---

[130] When you're watching a medical show and you hear them talking about getting a CBC, it means "complete blood count," which is an analysis of the quantity of these components in the blood.

[131] Literally, "red cells," which are what gives blood its red color.

[132] They call it an *anucleate* cell.

Hemoglobin

Hemoglobin is a protein—for the moment, let's content ourselves to say it's a big fat molecule. It's made up of four identical molecular chains arranged in a square—each is called a *heme*—and you guessed it, that's where the name hemoglobin comes from. One of the special things about a heme molecule is that, among other things, it has one metal atom—in hemoglobin's case, iron—in the middle of a ring of atoms called a *porphyrin*. This metal atom makes it possible to bind or attach to one diatomic gas molecule—in this case, O2. Thus a single hemoglobin molecule with its four hemes can bind four O2 molecules. Note the bond between the O2 and hemoglobin is strong enough for the oxygen to easily join and travel along with the hemoglobin, but not strong enough to prevent the next phenomenon from detaching the O2. When there's a lot of oxygen surrounding a hemoglobin molecule, the O2 will tend to stay attached. And when there's less oxygen around, the O2 is more likely to be released.

The hemoglobin in the red blood cells binds with the O2 from the lungs[133]—remember, oxygen has no problem at all getting though the cellular wall of the red blood cells. It happens spontaneously and automatically. The oxygen molecules are now passengers on the red blood cell.

Blood circulates in the body through an astounding piping system—the blood vessels—called the *cardiovascular system*: *cardio* for "heart" because the heart pumps to make the blood circulate, and *vascular* because it circulates in blood "vessels." The blood vessels are organized like a highway system: wide highways for traveling large distances, then a network of progressively smaller thoroughfares that lead to different places in the body. And when I say "astounding," I have carefully selected this word. Because we have between 75,000 and 93,000 miles of roads—yes, I said between seventy-five and ninety-three *thousand* miles of highways for our blood. That's long enough to go around the earth's equator more than three times! Inside every human body!

[133]   And technically speaking, it then becomes an oxyhemoglobin.

The human body—like any animal's that transports oxygen in the blood through a vascular system—over the course of many generations of evolution, has developed a clever solution to a complicated problem: how to move oxygen quickly from the lungs to the cells that need it, but move slowly enough that the oxygen can detach from the red blood cells when they are near those cells? Having a vascular network with progressively smaller vessels makes the cells slow down—like cars on a progressively narrower road, it finally gets so narrow a car can just get through; a vessel barely the width of the red blood cells is what we're talking about here. At this stage, you might be asking yourself where such complexity comes from—how have we come to have such complex machinery that is so complicated and so efficient at the same time—no spoiler here, but let's just say, since it came up . . . we still don't know the whole of the story.

So, we have red blood cells circulating in a very narrow blood vessel. They're traveling much more slowly than they were in the larger multilane arterial highways. The blood vessels are so thin that passing through their walls won't be a problem for the oxygen if it needs to do that. On the other side of the vessel wall, the cells—of the liver, the heart, a muscle, wherever—are bathed in a fluid called *lymph*. Lymph has a low oxygen concentration, so again—without any effort on our part—diffusion takes place, and removes the O2 from the hemoglobin. Then, the oxygen passes through the vessel wall into the lymph. And from the lymph, it again diffuses—once again requiring no effort on our part—into one of the cells. Let's use a muscle cell for this scenario.

The step when the oxygen separates from the hemoglobin is call *desaturation*.[134] Hemoglobin never gets completely desaturated. Some of the oxygen always stays stuck to its hemoglobin.

Upon entering the cell, oxygen binds to a protein that's very similar to hemoglobin—myoglobin, because we're visiting a muscle cell at the moment.

We find neuroglobins in nerve cells. In other kinds of cells, we

[134] In the medical field, oxygen saturation is a measure of oxygen concentration in the blood. You'll often hear "sat" for *saturation*.

Myoglobin

Myoglobin is a protein that, like hemoglobin, can bind oxygen, but unlike hemoglobin, it has only one heme and can bind only one O2 molecule. Here's another difference: Myoglobin's heme has a nucleus and a porphyn ring with an iron atom at its center. Here's another similarity: Myoglobin is red. The amount of myoglobin in an animal's muscle cells is the reason for the difference in appearance between red meat and white meat.

generally find cytoglobins. Their jobs are about the same.

What goes on next in the cell—and it's quite a process—is discussed in the next section (see page 209). For now, we'll simply say that by the time a cell has done what it's supposed to do with all this oxygen, various chemical reactions have produced a carbon dioxide molecule, CO2— made of one carbon atom plus two oxygen atoms. Then, because there's a higher concentration of CO2 inside the cell than inside the blood vessel, we have another round of diffusion that moves the CO2 molecules across the vessel wall into the slow-moving blood in the narrow,

And if They Don't Have Lungs?

Fish don't have lungs, they have gills. In some ways, gills are a lot like lungs. The membrane is folded on itself to fit into a confined space, and they have a large surface area for exchange with the outside environment. Instead of air being inhaled, water flows along the gills—because the fish is moving or because it pumps water with its mouth—and similar diffusion processes occur. The gill membrane lets in oxygen that's dissolved in the water and allows carbon dioxide to escape from the blood.

It's even simpler for insects. Air enters through a hole in the bug, then passes through tracheae (tubes) that divide into tracheoles (smaller tubes) that bring the air directly to the organs, and right there at the organs is where the cells exchange oxygen and carbon dioxide with the air.

one-lane blood vessel. A negligible portion of the CO_2 will dissolve into the plasma, and the rest will attach themselves, as polite CO_2 molecules should, to the available hemes of the hemoglobin in the red blood cells. Then they travel faster and faster, along larger and larger highways, until arrival at the alveoli in the lungs. This time, though, the blood contains more carbon dioxide than the air in the lungs, so "here we go again!" Diffusion causes the carbon dioxide molecules to move into the air in the lungs—outside the body. The hemoglobins, now free of the CO_2, are once again available to pick up O_2 and the cycle begins again. The only thing left to do is exhale to expel the CO_2-laden air from the lungs.

56. In a Cell

We are now among cells with proteins that have oxygen attached to them. So what's the oxygen for? Good question! To answer this question, we'll have to take a quick tour of a cell to understand what it is and what it does.

The cell is the smallest building block in what makes up the kingdom of living things. The smallest living thing consists of a single cell. The cell is to life as the atom is to matter—sort of anyway—and the analogy is valid because like that atom, the cell is composite and is made of other things. Its name comes from the Latin for a monk's cell, *cellula*, which is perfect because it's completely enclosed by a membrane, called a *plasma membrane*, inside which we find the protoplasm, which is made of either cytoplasm alone—the liquid, "gooey filling" of the cell— or cytoplasm and a nucleus, depending on whether the cell has a nucleus or not.

Cytoplasm—found in both eukaryotes and prokaryotes—is made of an aqueous solution containing mineral salts and various organic compounds. In eukaryotes, we find things called *organelles*, which are organic structures. We'll go into more detail about what they—some of them, anyway—do. Prokaryotes rarely contain organelles—but they do have ribosomes, plasmids—DNA—and a circular chromosome—that's

Prokaryotes and Eukaryotes

Prokaryotes and eukaryotes are two families[135] of living things that are distinguished by the fact that prokaryotes are cells that don't have a nucleus and eukaryotes are cells that do have a nucleus. Etymologically speaking, prokaryote means "before nucleus" and eukaryote means "having a nucleus." Using the traditional method for classifying families of living things, prokaryotes and eukaryotes are the very first distinguishing criteria—to distinguish the *six kingdoms of life.*

exceptionally linear . . . Got that straight? As is usually the case in biology, as soon as you dive into the details, you're in over your head! Needless to say, we are absolutely *not* going to get an in-depth look at biology in this book. My intent is more to satisfy your curiosity than make you feel like a complete ignoramus. I ask those who are keen on microbiology to forgive me.

Getting back to our O2 molecule hanging on to a myoglobin, a neuroglobin, or a cytoglobin, as the case may be. The oxygen is going to be delivered—yet again, completely spontaneously—to its final destination: a mitochondrion.

The mitochondrion—approximately one thousandth of a millimeter (0.001 mm) in length—is the cell's power plant. Within the mitochondrion, a chain of *oxidation reactions* occurs—these require oxygen, hence the name—with the goal of oxidizing carbon atoms to make carbon dioxide. This series of operations releases energy—chemically, by successively breaking up molecules and recombining their various elements—and the cell uses this energy to function.

Just so you know, a very long time ago, mitochondria used to be individual bacteria, completely separate creatures from our ancestors. But their role was so practical that they ended up living in perfect symbiosis within the cells where they could reproduce, and

---

[135]   Biologists, please calm down! I know I used the word *family* the wrong way when talking about the classification of living things, but again, folks, it's my book. No offense.

Krebs Cycle

The Krebs cycle is a whole set of oxidation reactions. It's an amazingly complicated process—by the way, all the steps occur, yes, spontaneously! The machinery is very well made. It starts with glucose that arrives at the mitochondrion the same way everything else does, by diffusion—glucose comes from digestion . . . from sugars and carbohydrates you eat. As mentioned, the Krebs cycle produces carbon dioxide and energy, but it also produces another molecule, *oxaloacetate*, a metabolite.[136] Oxaloacetate is the molecule that triggers the first step in the whole chain of reactions in the Krebs cycle—oxaloacetate restarts the whole thing! Like I said before, the machinery is very well made.

eventually, they became an integral part of us humans. Without them, we wouldn't exist! It's the same for *chloroplasts*[137] in plants. Come to think of it, it's fairly common for other organisms to support the human body's functions. How about this fun fact: A human being has one hundred trillion cells and carries ten times more bacterial cells— on your gums, in your saliva, in your intestines, etc.—to help your body function efficiently.

At this point, we've covered as many details as possible, without being experts in biology, about what respiration does. There's still a lot we haven't even touched on, such as this: How does the energy produced by the mitochondria become mechanical muscle power—say, for walking—and what goes on inside the cell's nucleus? Even if we consider only what goes on outside of the cell's nucleus, we could still spend whole pages on the *Golgi body*, *endoplasmic reticulum*, and *ribosomes*.

But if we really want to understand things more clearly, we have to keep our wits about us and radically change the scale to enter the most fantastic piece of equipment in the human body: the brain. It would be delusional to think we could discuss every aspect of the brain in a few sections of this book. We'd have to learn all the related anatomy,

---

[136] Small molecule produced by a living thing, thus coming from a metabolism.
[137] Performs photosynthesis in plant cells.

physiology, biochemistry, and neurology first. So we'll just focus on a few things that show how our brain is able to do amazing things, all on its own without any conscious desire on our part!

57. The Incredible Things Your Brain Does All by Itself

Before going into detail on this subject, we do need to have a little chat about the brain—more specifically, about the capacity of the human brain. People—the species, not the magazine—have this persistent idea that the human brain has some sort of phenomenal untapped potential that could almost make us superheroes. There's this idea that we are only a sneak preview of what the ultimate humans could be, merely a preliminary design concept of what the human race will, in time, become. It's the 10 percent idea.

That being said, more than sight, hearing, or speaking, memory is hands-down the most important tool the brain developed. Without it, barely any of the brain's faculties would be possible. And though we have a habit of talking about THE memory, we'll soon see that there are several, each as useful as the other.

Memory

The first kind of memory is *sensory memory*. It provides a sense of continuity with the present. Thanks to sensory memory, when someone is talking to you and speaks a word, you can still remember the beginning of the word by the time they finish pronouncing it. Without sensory memory, communication would be impossible. Words wouldn't even get to reach the areas of the brain that process language. It's called sensory memory because it's directly related to our senses and our perceptions of the outside world. The information is saved for an extremely short period of time, from a few hundredths of a second up to a maximum of two seconds. And among all the information that sensory memory continuously collects, and there's plenty, you have everything you see, including the painting on the wall you've seen from the corner of your eye for six years that you never pay any

10 Percent of Our Brain

The idea that we only use 10 percent of our brain is completely incorrect. Advanced medical imaging technology tells us, beyond the shadow of a doubt, that we definitely use our entire brain. And we have also learned that the least little brain injury can cause serious and irreversible damage. It is true, however, that we use only one part of our brain at any one time. Next you're thinking, perhaps, we could use our whole brain all the time, then we would be able to do things that seem impossible for us today, from thinking faster to telekinetic powers. Yeah . . . no. Our brain is approximately 2 percent of our mass, but it uses 20 percent of the energy we produce—up to 60 percent in infants under a year old. At any given moment, this energy is used to activate 1 to 15 percent of our neurons. It's likely that if we used all of our neurons at the same time, our brain would heat up enough to literally cook itself.

Albert Einstein is often credited with the 10 percent idea, but there's nothing to support that. Today, the idea is commonly used in science fiction to support the concept of a super human—sorry, Luc Besson.[138]

attention to; everything you hear, including the constant noise in the pipes that you never pay any attention to; everything you touch, all the time, without realizing it—your own clothes, the air, the back of your chair; everything you smell and everything that you taste at every moment—the air, your saliva; plus your balance, your sensing hot and cold over every square inch of your body! And, of that information collected every instant of every day by your sensory memory, only an infinitesimal portion of it is sent to your short-term memory; that's what we actually pay attention to.

[138]   Director of *Lucy*.

Sensory Overload

For a variety of reasons, the brain can be overwhelmed by too much sensory information at a given moment, causing a flood of nerve input. Depending on the individual and the nature of the overload, several reactions are possible. The responses can range from irritability to an epileptic seizure, with reactions such as panic or phobias showing up in between the two extremes. Overload can be provoked intentionally in a laboratory, making it possible to observe reactions such as fainting, overexcitement, aggressiveness, and hallucinations.

Sensory overload is never a pleasant experience. But it just goes to show that the brain is always carefully selecting the information to be processed. The brain is always on the job, and if it can't do it well, then as a minimum, it tries to protect us.

The selected information, that which our brain deems worthy of attention, goes into the *working memory*, aka the short-term memory. Short-term memory is a bit like RAM in a computer. It allows us to follow a conversation and it stores information that it needs to have readily accessible. And the same as with a computer, RAM uses more energy than turning a hard drive. Our short-term memory uses so much energy that it has to be limited. It's generally thought that we can store seven distinct items in our short-term memory. But that's not quite accurate. In reality, because each person is unique, on average we can store between five and nine unrelated pieces of information in our short-term memory at the same time. Perhaps you noticed I emphasized "unrelated" information. That's because when pieces of information are related to each other, they can be stored "in the same place" in our short-term memory. For example, the words *boat*, *car*, and *motorcycle* can be stashed in the same "box" in short-term memory, making it possible to store a lot more than seven pieces of information at one time, because the five to nine boxes can each hold a whole set of related information. Short-term memory uses three parts of the brain: the *central executive*, *phonological loop*, and the *visuospatial sketchpad*.

The *phonological loop* lets you temporarily save—short-term—verbal and sound information. When someone tells you a phone number and you don't have a pen to write it down right then, this loop makes it possible to hold on to the information. The life expectancy of information in the phonological loop is approximately two seconds. After those two seconds, the information is ready to be overwritten by new information that is now deemed by your brain to be immediately more important, whether you like it or not. That is why you need to repeat the number over and over while you are desperately searching for a pen to write it down. By repeating the number aloud, you're reactivating the loop as many times as needed.

The *visuospatial sketchpad* works a lot like the phonological loop, except as its name indicates, it's used primarily for storing visual information or information related to a spatial situation—in the sense of the space around you, not in the sense of "Space: The final frontier."[139] It's your visuospatial sketchpad that lets you follow something with your eyes, even if the thing temporarily leaves your field of vision. It also helps you build mental images, so it does an awful lot of work—just like the phonological loop, mentioned earlier—with the short-term memory. Suppose I asked you to describe the physical appearance of someone you know well, say a parent or a friend, but who isn't standing there with you. You'd construct a mental image of them in your brain. You'd search for visual information about the person in your long-term memory, and you'd reconstruct a composite image on your visuospatial sketchpad. This can't be done without the assistance of the third component of the short-term memory.

The *central executive*'s job is as easy to describe as it is complicated to explain: The central executive organizes information coming from the phonological loop and the visuospatial sketchpad, and it keeps necessary information available in long-term memory so that your short-term memory can work efficiently. How does it do that? That is still a mystery. Scientists can tell us which areas of the brain are assigned to the task. Scientists can tell when it's working and when it's

---

[139] "Space: the final frontier," James Tiberius Kirk, *Star Trek*, 1966.

not working—that's simple: It's working all the time. But they still can't explain the central executive's mechanisms or its chemistry. Even today, most of the brain remains a vast and uncharted territory.

And at long last we come to *long-term memory*, the hard drive of the brain. Long-term memory is where memories are stored, whether from a few hours ago or from many years ago. Anything you can think about that occurred more than two seconds ago is stored in your long-term memory. It works in two different ways, in parallel, depending on whether the memory you wish to bring up is declarative memory, aka *explicit* memory, or nondeclarative memory, aka *implicit* memory. As its name indicates, explicit memory is what you explicitly call up. Such is the case with things you learned associated with a date. When you say to yourself that "1776 was when the Declaration of Independence was signed," it's in your explicit memory. It's the same for sensory information, of course. If you mentally search for a smell or a tune, it's in your explicit memory. Watch out, though: Sometimes you can bring up a memory in an explicit manner without realizing it! I understand this seems like a kooky thing to say, but allow me to explain. Imagine you're in a room at a friend's house, and a fragrance suddenly takes you back to your first kiss. You don't think you thought of bringing up this memory, but that's what happened. Your nose instructed your brain—or more precisely, your perception of the perfume's smell triggered your short-term memory—to search for a matching smell in long-term memory. The

Memories and Perceptions

When that kind of memory explodes in your head, sometimes it feels like you're reliving the memory, or almost anyway. It's mind-boggling when you realize we experience events only through our senses. So consequently, if those perceptions are reproduced in the brain, it is theoretically impossible to distinguish the memory from reality. In a more limited way, hearing a song in your head when you're trying to think of the title or the lyrics activates the same areas in the brain as when you really hear the song. Aside from the fact that the auditory nerves aren't in use, there's no difference.

intensity of the memory—your first kiss—left the door wide open for this memory, which was actually explicitly "called up" without your specifically having asked for it.

Explicit memory storage is distributed over multiple areas of the brain. There's the frontal lobe, where learned knowledge is stored; the temporal lobe, which stores factual information; and various other areas linked to the senses for *episodic* memories—related to real, dated memories. And let's not forget the visual area for images, plus the auditory cortex for sounds, etc. We also have the hippocampus, which is the area of the brain that converts events into memories, and the frontal lobe, which checks to make sure the memory is real and not a fantasy.

Procedural memory comes under implicit memory, and it's different. It's what you might consider *unconscious* memory. Procedural memory is why we have the saying "It's like riding a bike." Cycling is actually a very good example, because once you know it, you know it for good. And even if you haven't ridden a bike for twenty years—and I speak from experience—all you have to do is get back on a bike and after a few minutes, you're a pro again. But riding a bike isn't simply about pedaling. It involves continuously maintaining your balance, looking around to watch where you're going, subtle body movements to shift your weight forward or a little to the right, as well as coordinating all these things that need to be done. You never even think about it. You just ride the bike.

But once upon a time, you did actually learn to ride the bike. That learning was stored in your procedural memory. Habits are also stored in procedural memory—I'm sure you have the following experience with going somewhere you routinely go: Sometimes you go on autopilot and automatically follow your usual path and you arrive home without really thinking about it. And sometimes smokers automatically stub out a cigarette before it's even lit! Implicit memory is stored in three main places in the brain: first there's the *cerebellum*, which is in charge of coordinating movements; next, the *caudate nucleus*, in charge of instinctive actions; and finally, the *putamen*, which stores acquired skills, like being able to ride a bike or knowing how to swim.

For a motion or an action to be recorded in procedural memory, you have to repeat the motion. Repetition allows it to become a habit—automatic and unconscious—which means the brain has created permanent neural pathways for performing the motion in order to accomplish it using the smallest amount of energy possible. So the motion becomes automatic. This is great for a violinist or a golfer, but a lot less great for someone who smokes cigarettes or bites their nails. Don't be fooled, though: The habitual motion isn't the only reason it's so hard to quit smoking.

Ever notice that sometimes your explicit memory can't bring up a movie title, an actor's name, or maybe that synonym for *constraint* that you are absolutely sure begins with "re"? So you ask everyone around you, and everyone can come up with *constraint*, and other synonyms or equivalent expressions; but no one can come up with that darned word *restraint* that you were really looking for—and, anyway, it isn't even a synonym of *constraint*! We call that having a word on "the tip of your tongue."

On the Tip of Your Tongue

I want you to understand that having a word on the tip of your tongue is an exceptional phenomenon. We'll need to go over it for you to understand that. Think about it: Every word you pronounce—in fact, every word you even think about—is the result of a mechanism that's as complicated as it is efficient. This mechanism draws words from the vocabulary in your memory and helps you make sense of it all using a set of intricately detailed rules that we call language. It's running nearly all the time. It's truly exceptional to hit a hiccup, especially when you consider the number of words it handles without a hitch.

So what's going on when it hits a glitch? First of all, and this shouldn't be a surprise, it frequently happens with words you don't use very often. They are literally buried in your long-term memory, and digging them up can be quite a feat. On the rare occasion when you search for a word and can't think of it, the real problem is that other words take its place in the phonological loop. Perhaps because they start with the same sound, even though they don't mean the same thing at all, or maybe because they are in the same lexical field,[140] or simply because your personal experience created links between the word you're looking for and others.

Once your phonological loop is full, no other word can be copied to it unless you free up space. And while you're desperately searching for this word, you try to find it among the other words that are already located in the phonological loop, which are there because they resemble it, or mean the same thing, or whatever. Trying to use the other words to find the word you want is actually the problem. The phonological loop is full and you repeatedly stimulate the contents; thus they remain active and don't free up any space for the word you're looking for. Fun fact: When you're searching for a word and you ask those around you— your friends, your family—it is quite common for them to become tongue-tied as well. Evidently, it is extremely contagious!

To succeed in finding *the* word, there are two strategies that gener- ally work: The first strategy consists of allowing your mind to wander and pronouncing each word that comes to mind, allowing your brain to freely associate words with whatever it wants. For the second strategy to work, you have to free up the phonological loop—that's the reason the word suddenly pops up as soon as you give up looking for it; sometimes it's only a few seconds after you say "Oh, forget it!" and you stop trying to think of it—your phonological loop purges itself, and bingo!

Oh, and speaking of contagious, we have to talk about yawning, because in case you didn't already know, yawning is contagious. Have you ever noticed that when you see someone yawn, you start to yawn? In fact, just reading this sentence might make you want yawn—but you're just going to suck it up and wait. Yawning is yet to come (see page 238)!

Inner Speech

Remember a few lines ago, I said *thinking* about a song activated the same parts of your brain as *singing* the song? Guess what—it's the same with inner speech. As you know, when you're thinking, you hear your own voice in your head, pronouncing the words you're thinking. Studies

[140]   Words related to the same subject.

have shown that this voice is as real as it could possibly be. In fact, when you think, your brain activates the same parts of your brain as when you actually speak your thoughts and "think out loud." Here's what's going on: In a way, your brain anticipates the sound of your voice and makes you hear it in your head. Other than the auditory nerves not registering hearing your voice, everything else is exactly as if the words were really being perceived audibly by your ears. If, as Hegel says, "it is in words that we think,"[141] shouldn't we ask ourselves how someone who is deaf or mute from birth thinks, because they've never heard their own voice? And of course, we know today—unfortunately, that hasn't always been the case[142]—that those who are deaf or mute are just as able to think as anyone else.

Deaf or Mute from Birth

Those who are deaf from birth have to develop other ways of thinking, obviously, since words are completely inaudible to them. Studies have shown that those who are deaf think visually, whether it's simply visualizing images, or in some cases written words, or if they learn sign language early enough for it to be their first language, they think in sign language.

For those who are mute, there are two solutions. Either they develop visual thinking, like those who are deaf, or a substitute voice replaces their own missing voice, usually the voice of a mother or father figure. Brain activity studies have been performed to see which areas of the brain are active during thought, and they clearly showed what kind of thinking the person was using.

Our brain is very good at doing what's needed to help us. It's always trying to connect back to a typical or familiar situation. That's why we think audibly or visually in our head. But sometimes, the brain reaches even further back to rely on what were originally basic survival instincts.

---

[141]   Georg Hegel, *Encyclopedia of the Philosophical Science in Basic Outline*, pt. 3: *Philosophy of the Spirit* (1817).

[142]   The word *dumb* can mean "mute" or "stupid."

Face Detection and Facial Recognition

We've already briefly covered pareidolia (see page 125), the illusion that makes us see the shape of familiar objects—like animals, humans, and faces where there aren't any—in plants, clouds, cars, etc. When it comes to faces, our brain has two powerful and effective tools—so effective that if they stop working properly, it becomes immediately evident how much we rely on them. These tools are *face detection* and *facial recognition*.

In humans and primates, detecting faces is an ancient primordial skill, much of it innate. From birth, infants can detect a face—note there is a difference between *detecting* a face, meaning knowing it's a face, and *recognizing* a face, meaning knowing who it is. This ability is the result of *evolutionary pressure*, or more precisely, *biotic pressure*.

Biotic Pressure

Biotic pressure a natural phenomenon that drives evolution of a species subjected to a stress exerted by other living organisms, whether the same species or not. To survive, our early prehuman ancestors had to learn to quickly identify any kind of possible danger, be it a predator or an aggressive prehuman.

There are also *abiotic pressures*, which are more associated with environmental conditions, like living in a polar region or, the opposite extreme, in a desert area.

Many areas of the brain are involved in face detection, but in 1986, Dame Victoria Geraldine Bruce,[143] a British psychologist often known as Vicki Bruce, and Andrew Young, a British psychologist often known as Andy Young, developed a theoretical model describing the cognitive processes in face detection. This model is widely known as the Bruce and Young model.

---

[143] No connection.

Cognitive Process

The brain is an incredibly complex machine, and many of its mechanisms are still very much a mystery to us. But there is a scientific way to simplify its operation. The same method is used in computer programing; we call it encapsulation. In computer science, encapsulation consists of *encapsulating*—hence the name—an unnecessary complication behind an interface. Starting a car is a classic example. I'm sure you suspect that some relatively complicated things happen when you turn the key in the ignition.

Fortunately, understanding the complications of getting the engine running using a process that involves internal combustion—or for getting electrons flowing in a controlled manner through the motors in the wheels of an electric car—isn't necessary for being able to drive the vehicle. The complicated process is completely masked behind a simple interface: turning a key. In the same way, thanks to computer science, you are completely unaware that hitting the key "A" on your keyboard triggers numerous operations—literally hundreds—before an "A" appears on your screen; this complexity is masked by the interface called the *keyboard*.

Cognitive processes are a bit like these simple interfaces when it comes to how the brain works. They give us a simplified way to discuss what's going on. When you hear a sound, your brain activates a phenomenal number of mechanisms to perceive, and then identify, the sound. Suppose you're doing a study: Depending on its scope, you can choose to ignore all the complexity, and simply choose to define the process as having three major steps: sensation, perception, and finally recognition of the sound. These—super simplified, here—steps are exactly what is meant by cognitive processes.

Bruce and Young wanted to describe the general steps of how a brain detected—then recognized, that was part of the study, too—a face. And here are their conclusions: Detection and recognition of a face occurs in seven major steps (the internal operations of these steps are still, to this day, not understood). The five first steps are for detection of a face and the last three are for recognition. Yes, I'm aware that

that sounds like eight steps, but the fact is, the fifth step, as we shall soon see, straddles both detection and recognition. It's this step that determines if there is a need to start the recognition process or not—which is really a good thing.

The first step, called *pictorial encoding*, consists of analyzing purely visual information, such as contrast, brightness, color, etc., to determine if a coherent shape emerges from the image perceived by your eyes and to determine if this shape has the general characteristics of a face. The brain brings up various generic representations of what a face is, depending on the view from the side, from three-quarters, from above, or from below. It's in this step that your brain can make you see a smiling face in the headlights and grill of a car; this is also the step where a circle, two dots, and a curved line can resemble two eyes and a smiling mouth, a smiley face. This step is always on the job, regardless of what you're looking at. To your brain, it's a matter of survival. As you read each word on this page, the image formed is run through this face-detecting machinery to see if there's a face in there somewhere.

The second step, called *structural encoding*, extracts the invariants of the face: position and distance between the eyes, shape of the nose, etc. This step, without regard for the expression on the detected face, forms a mental 3D representation of the face. At this point, in a way, face detection is complete and ends. But your brain continues on. And even before it tries to recognize the face, it wants to know if there's anything to be worried about. For the first time in the process, it pays attention to the context of your perception.

The third step is *expression analysis coding*. This is when your brain goes through its catalog of known facial expressions in an effort to understand the emotional state of the detected face so it can quickly determine if the face looks aggressive, frightened, sad, happy, serene, etc. This step is primordial, because depending on the determination, if the brain sees a danger, it can start an emergency procedure, which is quite cleverly called the *fight or flight response*.

Fight or Flight

In 1929, Walter Bradford Cannon, an American psychologist, developed a model for describing an animal's response—and in particular a human animal—when faced with a direct and imminent threat. His rather simplistic model has since been updated, but it remains a point of reference for the subject. When there's an immediate danger, the body prepares for mainly two possible responses: flee from the danger or fight it. To this end, regardless of the response used, several mechanisms are triggered automatically, such as secretion of *catecholamines*—adrenaline, for example—to facilitate preparation for sudden muscle movement and the expenditure of energy that goes with it.

This energy will require oxygen, so your heartbeat goes up, and there's a decrease, nearly a complete shutdown, of the digestive process, so the maximum amount of energy can be directed to the muscles. At this point, your breathing accelerates; you breathe harder and start to feel nauseous; you get "butterflies in your stomach." Except for the blood vessels serving your muscles and vital organs, your blood vessels constrict in multiple parts of your body to allow the blood be diverted more efficiently to the muscles. The increase in your heart rate heats you up, so you sweat. You also get goose bumps to help regulate your temperature—while all at that same time your fingers, toes, and nose get colder, your body contracts into a position for fight or flight, and trying to fight this contraction makes you shake. Last, your hearing becomes less sensitive, and so does your peripheral vision to increase your forward vision: You get *tunnel vision*.

All this happens whether you like it or not via the *sympathetic nervous system*. There are three major parts to your nervous system: the sympathetic nervous system, the enteric nervous system, and the parasympathetic nervous system. The sympathetic nervous system controls the unconscious activities of your body, such as your heart rate; the enteric nervous system controls the digestive system; and the parasympathetic nervous system, along with the sympathetic nervous system, controls involuntary activities of your organs.

And knowing whether to fight or flee—I almost enjoy telling you, "It's not up to you." Depending on the situation, you brain makes the choice for you. In dangerous jobs—military, special response teams, police, firefighters, etc.—special training is required to get the brain accustomed to stressful situations.

The fourth step is *facial speech coding*. It helps you understand what a person is saying through the movements of their mouth and face. This step is also involved in the analysis your brain makes to evaluate a possible danger.

Dubbing

When a movie is translated into another language, there are two possible strategies: dubbing or subtitles. The attractiveness of subtitling is that the original soundtrack of the film remains intact, but it has the major inconvenience of dialogue's being dense and rapid: The subtitles have to be simplified to keep up with the pace of the film, because we don't read as fast as we listen. Dubbing consists of having translated phrases pronounced by other actors and replacing the original voices with dubbed voices. But this creates a *cognitive dissonance*: The brain has to process two pieces of contradicting information—we'll talk about that soon, but just for the record, the brain does not like contradictions. The sounds of the dubbed words don't match the lip movement of the actor on the screen. In some cultures, such as France, they are so used to dubbing that the dissonance is barely noticed anymore, but only because dubbing companies do a lot of research to get sentence lengths and sounds that closely match the way the original speech looks! Except for advertisements—lower budget—sometimes the obvious difference between what you see and what you hear is shocking. A trick that's often used in documentaries—they don't want to spend half their budget on dubbing—is to make the voice of the original speaker faintly audible in the background when the dubbed voice speaks.

The fifth step is a pivotal step between detecting and recognizing a face: *directed visual processing*, aka *visually derived semantic coding*. We can break this into two parts: First is *semantic information coding* from visual processing, which consists of assigning the perceived face with a set of general contextual signals such as age, gender, ethnic origin, etc. The two preceding steps and the first phase of this step will possibly activate a *facial recognition unit* in long-term memory to search for a resemblance between all the processed information— visual and otherwise—and a familiar face. The second part of this step is the *facial recognition unit*. It activates yet two more facial recognition steps.

The sixth step is a step to process the person's identity, from a semantic perspective: *person identity nodes*, aka *identity-specific semantic coding*. This step is different from the preceding, in that the information being processed is stripped of the current context in which the face is perceived. For example, let's just say you're on vacation at the beach, and you walk by the guy who usually serves up your kebab at the corner kebab cart. Your brain might make you remember the taste of your kebab. The associations seem completely random because they are so inconsistent. But they place a perceived face in a remembered, familiar context from the past.

And finally, the seventh step, *name generation*, which could be considered as formal identification, is the step during which your brain has recognized the face and searches in your long-term memory for the name of this person as well as any biographical information you may have about them. At this point, identification is completely finished. The next time you see a close friend or relative, notice how the whole identification process for this person seems instantaneous. It gives you a good example of your brain's astounding speed and power.

In 1988, Andy Young and another British psychologist, Andy Ellis, developed a new model specifically for facial recognition. It was based on their study of two different pathological syndromes associated with facial recognition, *prosopagnosia* and the *Capgras syndrome*.

Prosopagnosia and the Capgras Syndrome

According to Ellis and Young, once a face is deemed to be a face, two parallel processes operate at the same time to identify the face in question: *conscious facial recognition* and *affective* (or emotional) *facial recognition*, also known as *overt recognition* and *covert recognition*, respectively. As their names indicate, overt facial recognition is the process in which your brain searches in your long-term memory for factual information, related to this face's recognized features, such as name, biographical items, and mental images of the person. Affective or covert recognition serves to remember the emotional connection to this person. A very strong emotional connection for a close relative, and less intense emotional connection for a random acquaintance. Now this is interesting: If there's someone you barely know, but met under positive conditions—on vacation, in good company, etc.—that person can evoke a stronger emotional connection than a person you know a little better but for whom there are no pleasant associations. And because these two processes occur in parallel, brain injury or a cognitive issue can cause two possible ways to *not* recognize someone.

Prosopagnosia[144] is not being able to recognize a face—affected individuals can see the features of the face, they are completely aware that it's a face, they just cannot recognize the person in question. That means the formal-recognition-of-faces step isn't working. A person with prosopagnosia could be sitting with their own parents and not recognize them except by other clues—context, voice, etc. This problem can be innate and exist from birth—that's congenital prosopagnosia—or it can be acquired, occurring after a brain injury. CVAs[145] are responsible for 40 percent of the acquired cases.

It's also possible to have failure in affective facial recognition—remember, the emotional response to a face. It causes a very unusual problem, and sadly, it's frequently accompanied by psychiatric disorders—and you'll see how it must be very hard to tell whether or not the disorders

[144]   From the Greek, πρόσωπον (*prosopon*), which means "face," and ἀγνωσία (*agnosia*), which means "not knowing" or "ignorance."

[145]   Cerebrovascular accident.

are simply caused by the first problem. When affective facial recognition isn't working properly, you can indeed recognize people who are close to you—your parents, for example—but you feel no emotional connection with them at all. This causes a cognitive dissonance. Your brain resolves the issue by deciding that your loved ones are actually impostors. The impostors look just like your parents—formal recognition is working fine—and you have absolutely no emotional connection with them—so it *seems* like the affective recognition is working, too. Thus your brain doesn't see a contradiction anymore and is "content."[146] This disorder is called *Capgras syndrome*, or *Capgras delusion*, named for the French psychiatrist Joseph Capgras, the first to describe it clinically in 1923. This disorder, whether caused by a brain injury or a brain malformation, generally cannot be differentiated from mental disorders called *chronic nondissociative psychoses*. In fact, an affected patient is completely convinced that the people she knows—all the people she knows—have been replaced by identical impostors. Can you imagine the distress experienced by someone who is trying to explain to the psychiatrist that their own parents are merely impostors, and all the while she might be thinking the psychiatrist has been replaced by an impostor, too?

There is an even more troubling disorder than the Capgras syndrome. It's an extension of the Capgras syndrome, accentuated with a profound paranoia: *Fregoli delusion*.[147] It's nearly impossible to tell whether this rare disorder is a structural problem in the brain or a mental illness. As with Capgras, the patient is convinced that all the people he knows have been replaced by impostors, but with a difference: There's only one impostor who disguises himself as different people the patient knows. The mere fact that the brain is capable of connecting multiple people to a single individual when the faces are formally recognized as looking different is an immediate clue as to the scale of the

---

146   I've given the brain a personality for purposes of illustration. It doesn't have a personality. The brain is an organ of the body, just like a lung or the liver.

147   Named for Leopoldo Fregoli, the famous Italian master of disguise and quick-change artist at the end of the nineteenth and beginning of the twentieth centuries.

problem. In the Fregoli delusion, the others are always the same one imposter.

Vision, as you'll soon see, is not a simple mechanism. Vision is the result of three billion years of evolution, and you'll be surprised by what vision can do, especially when you hear about blind people who have *blindsight*.

Blindsight

You've probably seen in a movie or heard about a blind person who can *sense* if she is alone in a room, avoid an obstacle without touching it, or instinctively tell if a person in front of them is angry, without a single word being said. Blind people themselves don't know how to explain it. But many of them think it's because their other senses are always on alert, to the point that they can hear another person in the room breathing or are able to use changes in reflected sounds to avoid obstacles. Nope. It's none of that. It has more to do with the fact that there's blind, and then there's *blind*.

Let's back up a bit and review how we see with our eyes. When an image is perceived by the eye, it's sent in as electric impulses via the optic nerve to the *primary visual cortex* located in the back of the skull. The primary visual cortex collects information from the two eyes and builds a three-dimensional image. But what I didn't mention last time was that before the information gets to the famous primary visual cortex, it first takes a short trip via the *superior colliculus*—in the middle of the brain, above the thalamus—which is mainly responsible for direction of gaze and, most important here, is an unconscious part of the brain.

Some blind people have eyeballs that don't function—meaning, physically, the eye itself has a problem with the retina or optic nerve— while others are blind due to a problem in the primary visual cortex. Those in the second category are truly blind, of course. They don't have some psychological block; they are blind. Period. The only difference is that their eyeballs work properly. Their eyes perceive light and send information to the primary visual cortex. And this means the trip through the superior colliculus happens as it normally would. Thus this unconscious area of the brain still senses images sent from the eyes, allowing the blind person, who can't consciously see anything, to unconsciously feel a sense of danger. And, I might add, this trip through the superior colliculus is probably what causes the feeling of *déjà vu*.

Déjà Vu

No one knows exactly what déjà vu is, mainly because it's extremely difficult to reproduce it in a laboratory while you have someone comfortably (yeah right) situated in an fMRI.[148] It's so difficult to reproduce that we don't even know if it's possible to reproduce it. So, about déjà vu, there are several hypotheses to explain it. But first, for those of you who don't know what déjà vu is, it's an extremely strong feeling while you're experiencing an event, even a normal everyday event—a cashier giving you 60 cents change, saying "and that makes five"—of having already lived that exact moment. Not just a similar moment, or an identical moment—no, that exact same moment. After a few seconds, the sensation completely goes away. But while the sensation lasts, you'd bet your right arm it was real.

As I said, there are several theories for explaining this common phenomenon—not "common" in the sense that it happens every five minutes, but "common" in the sense that it happens to everyone—I will now gently eliminate a few theories while trying not to offend anyone: memories of a previous life, seeing your life in an "alternate universe," a premonition, all that kind of stuff . . . I'm going to say right here, they're all based on absolutely nothing—these are no more valid than saying déjà vu is caused by dogs that speak Finnish when no one can hear them. I certainly respect the beliefs of others, whoever they may be, as long as they don't require me to accept their beliefs, but these theories are quietly going up onto the shelf with summer clothes and Grandma's Ouija board.

In all probability, the feeling that you repeated an instant in time was an illusion created by the brain to remedy a contradiction in its perception of time. In other words, when the brain takes in information, it

---

[148] Functional magnetic resonance imaging.

does so sequentially, processing so many images, sounds, or other data per second. So if the same data arrive twice in succession, but over too large of a time span to be considered simultaneous, your brain is completely willing and able to sweep the problem under the rug by telling you that it wasn't simultaneous at all. It tells you instead that it was actually two separate events. First, let's consider a visual example. As mentioned, eyes sense light at the back of the retina, then the light is transformed into electrical nerve information via the optic nerve; this information goes through the superior colliculus and ends up in the primary visual cortex, where the perception of a 3D image is formed. Yes, in 3D, because there are two eyes receiving that light, and each eye has its own separate nerve pathway from the optic nerve all the way to the primary visual cortex by way of the superior colliculus—by the way, the pathways cross over in the brain, but we don't care about that here.

One of the theories says: For some reason, in some situations, the information from one eye lags behind the information from the other eye. It's ever so slightly delayed—on the order of a hundredth of a second. When the brain receives information coming from one eye, then the other, visual continuity drives the brain to correctly transmit the perception of this continuity, and at the same time raises the issue of concomitance: Are these two pieces of information simultaneous? When in doubt, the brain will say yes *and* no, which gives you the sensation of continuity of action *and* replication of the instant. Once this event leaves short-term memory, the brain cleans it up by saving only a single occurrence, which is why we often speak of having a déjà vu, but we don't normally talk about feeling a déjà vu.

Another theory relates to the superior colliculus. Some researchers think that under certain unknown conditions, the superior colliculus may successively resend, in rapid succession, a piece of information it processed. So according to this theory, two identical pieces of information indeed arrive at the primary visual cortex one after the other, and so "Here we go again!"—the brain does what it can to maintain the continuity of the action while preserving the two separate pieces of information. And one thing we must not forget about déjà vu: To the brain's short-term memory, it's not a replication of an instant that just happened—the first occurrence isn't in memory anymore—the brain resolved the problem using some good ol' sleight of hand, letting you think "yes, this really is a duplicate of an instant I already lived through," your brain tells you, "but it was a very long time ago, it's buried in the long-term memory, no need to dig any further, everything is fine, it will all be forgotten in a few seconds anyway." There it is in a nutshell. Does that work for you?

In the simplest terms: Two images arrive at the primary visual cortex, one after the other, even though they are images of the same instant: The brain tries to get away with saying, "Well, they're two successive events!" to which all the other parts of the brain that handle the other senses respond, "I don't think so, Tim! There was definitely no repetition of the exchange of money"—going back to the example of the cashier. The primary visual cortex, quite annoyed, ends up negotiating with the other parts of the brain: "What if I tell you it's just a memory? Does that make you happy?" And then, of course, some smart-aleck part of the brain says, "Hey, there's no sign of it in the short-term memory," then the visual cortex comes up with, "Right, because it's an *old* memory!!!" and that one works every time.

The brain does not like paradoxes and it doesn't like gray areas, either. You just saw a little of that with the déjà vu. The brain puts a great deal of significance on maintaining continuity of time. But have you ever noticed that during an accident or in a stressful event time seems to happen in slow motion?

Time in Slow Motion

In stressful situations, like an accident or a fistfight, besides the flight or fight response, sometimes you also get the impression that time slows down. During an accident, it's particularly frustrating, because time seems to slow down, but you don't seem to be able to react at normal speed, which would be faster than the event is taking place. Remember what I said a few lines above, about the brain processing a given amount of information per second? Your perception of time passing is based on this speed. And this speed, even though it's always varying slightly, *seems* normal. However, when a particularly stressful event suddenly occurs, your brain races and starts to process a lot more information, a lot faster, but not all types of information—as mentioned earlier, hearing is reduced and there's a partial loss of peripheral vision (see page 224). To put it another way, we could say it reduces the number of information sources, while preserving the overall density of information taken in. We could simplify this by supposing it senses half as much data from peripheral vision and twice as much visual data from straight ahead— this simplification isn't correct, but at least you get the idea of what we're talking about.

High-Speed Camera

A standard video camera records between twenty-four and thirty images per second, depending on whether it's for theater or television, and even on what part of the world the film is for. By showing these images at that same speed, we watch a movie that runs at normal speed, which is what we're used to. Every second that we personally perceive as we watch corresponds to 1 second in the film. Some cameras can film things in slow motion; these cameras record a lot more images per second—some can record tens of thousands of images per second. When we watch film at a speed of twenty-four, twenty-five, or thirty images per second, every second that we personally perceive as we watch corresponds to a fraction of a recorded second. That's how they film action in slow motion.

Let's agree that your brain typically takes in 90 straight-ahead images per second. During an accident, that rate increases—let's say to 180 images per second, just to give ourselves a round number to work with—so your perception of the event, which doesn't change and continues running through actions at 90 images per second, will be that time seems to pass at half of normal speed. It's only an illusion in your brain. Time always passes at the same speed—generally, at our scale . . . we'll see with special relativity that this is only a general approximation[*]—anyway, that's why you can't react any faster. The event still happens at normal speed; it's just that your perception seems to indicate otherwise.

Would you like to hear another example of your brain playing crazy and sometimes dangerous tricks on you?

High Place Phenomenon

Not everyone experiences high place phenomenon (HPP). That's a fact. But for those who suffer from HPP, the process is always generally the same. You're at the top of a skyscraper, the edge of a cliff, on a balcony, at the top of a ladder, standing on a table—seriously, if you have HPP, please start by not doing those things—and a strange feeling comes over you, somewhere deep inside . . . you feel the sudden urge to jump. Or in some cases, you feel like someone is pushing you . . . even though you're alone. Here again, the mystery can be solved pretty easily: Your brain is processing contradictions as best it can.

So then, what's the problem? On one hand, your brain is in red-alert mode; because you're at the edge of a cliff, you are literally close to death. But on the other hand, you feel completely balanced and stable on your two feet; there's no particular risk of falling. Ordinarily, you don't fall for no reason, so in reality, there is no danger. However, your survival instinct—which is much stronger than your senses—is freaking out because you are risking certain death. Your brain *never* jokes around when life and death are on the line.

And thus, again your brain is confronted with a cognitive disso-nance, a contradiction between two pieces of information: on one hand, the perceived stability of the situation, and on the other, the menace of death. So your brain resolves the problem as best it can. Because there's no particular reason to fall, it decides, "Someone *is* pushing me!!! See I'm right!" Or maybe, "I actually want to jump. Oh! Oh! That's danger-ous too!!! Your brain isn't actually having a conversation with itself, this is just a dramatization made especially for you.[149] To your brain, these are the only possible explanations for fearing this immediate danger of death. For those who don't experience HPP, the threat of death seems

[149]   Because I care.

less immediate, and the brain can calm itself and remain in its comfort zone. The brains says, "Yes, I'm sure there's actually nothing here to make me fall." That's the difference between people who are sensitive to HPP and those who aren't.

Sometimes, the perceptions from our senses aren't interpreted correctly. There's a whole family of disorders called agnosia. And we're going focus on a particularly spectacular one, *hemineglect*.

Agnosia and Hemineglect

Now, on to hemineglect, aka spatial neglect, as well as unilateral neglect and hemispatial neglect. As we saw when we talked about prosopagnosia, the term *agnosia* means "ignorance"—so, prosopagnosia is an illustrious member of the larger family agnosia. Sometimes the person's senses may be perfectly functional, but the perception isn't processed—it's as if the perception didn't happen. The syndrome may be more or less debilitating depending on which sense is affected. Thus you can have someone who isn't deaf, who understands the words they hear, and who can hold normal conversations, but they are incapable of telling the difference between nonverbal sounds. A fire siren, a rock plopping into the water, and a phone's ring tone are all the same to them. They can hear the sounds well enough, they just can't tell them apart. It's called *auditory agnosia*, or more specifically, environmental sound agnosia. Along the same lines, the inverse is possible: A person can distinguish all the sounds they hear, but they can't recognize a single word. This is called *pure word deafness*, or auditory verbal agnosia. Other types of pure word deafness might allow one to converse and distinguish sounds, but not music—I don't mean recognizing a song and calling it up from memory, I mean music can't be recognized as music. In this case, it's called *amusia*—we might be able to make a number of puns here, but none is funny.

Cerebral achromatopsia is difficulty in recognizing or telling the difference between colors—color blindness that originates in the brain. It should not be confused with color blindness (aka Daltonism), which is the function of a genetic anomaly in the eye's cones. In the more complex *associative color agnosia*, a person can distinguish color perfectly but cannot associate colors with an object. You can give them four different-colored plastic blocks, and if you ask them to pick up the blue block, they can't do it. Ask them the color of a cucumber or a banana, and they can't answer you.

Hemiasomatognosia is characterized by not being able to recognize half of one's own body—right or left depending on whether the problem is in the left or right side of the brain—even though that half of the body is perfectly functional, without any sensory or motor problems. These individuals often feel like someone else is in bed with them, even though they're alone. This disorder is different from the *alien hand syndrome*, also called *Dr. Strangelove syndrome*.

Dr. Strangelove Syndrome

Individuals with Dr. Strangelove syndrome have an appendage—usually an arm—that seems to act completely on its own, and the individual isn't able to control it. Imagine, if you will, that as you are buttoning your shirt with your right hand, your left hand is unbuttoning it and you can't do anything about it! Or one hand throws away the cigarette that the other hand just put in your mouth—certainly cuts down on your smoking, anyway. Because the affected person cannot recognize these movements as movements they can control, they have no sense of ownership. So, it's fairly common for someone to personify the alien appendage and have a nickname for it.

Hemiasomatognosia is a form of hemineglect—which ignores half the body—but there are some amazing kinds of hemineglect in the family of visual agnosias. The simplest one to understand is that of unilateral spatial neglect; the eyes seem to receive only half of the information. It seems like only one eye is working properly—in reality, both eyes are working normally, but due to a problem in the primary visual cortex, the information from one eye is lost. Obviously, this causes problems for anything that has to do with three dimensions—especially for depth perception—but what is perceived by the eyes is perceived normally, though only partially.

And the craziest of all visual-related hemineglect is *object-centered neglect*. The aforementioned is the most mind-blowing spatial agnosia there is. I want to be clear here: I may seem to find this captivating, and that is definitely the case, but I want you to understand that I have much sympathy for these folks. This spatial agnosia must be a nightmare to live with. Imagine everything looks normal on the right and on the left. But as soon as you focus your attention on someone or something in particular, you can only see half of it—let's say the right half. Thus you talk with the right half of any individual; you watch the right half of the TV, but as soon as you focus on one of the actors, suddenly you can see only half. And here is what fascinates me the most with this hemineglect, and it shows just what kind of crazy things your brain can do: Suppose you are looking at someone who is facing you, and you can see only their right side. Then if they turn around with their back toward you—and this is where your brain is so completely badass—you still see the same side, but on the left now . . . and from the back!!! That's rather absolute proof that the problem isn't your eyes.

The worst situation, and it's fairly common, is to have one agnosia coupled with another agnosia, called *anosognosia*, which is characterized by being unaware of being affected by an agnosia.

The Hypnic Jerk, or Sleep Start: Hypnagogic or Hypnopompic

The last point I want to talk about regarding the brain is a phenomenon that's not rare or popular, and we've pretty much all experienced it at one time or another, usually before or during adolescence. Hypnic jerk, or sleep start, is the sensation of suddenly falling when you are just about to fall asleep—in this case, that's *hypnagogic*—or the opposite, and more rare, just before you wake up, in which case it's called *hypnopompic*.

When you fall asleep, you typically go through several steps as part of the process of falling asleep. But sometimes due to stress, fatigue, or any number of reasons related to your body or your surroundings, you fall asleep too quickly without going through all the steps. And with that, your brain . . . starts to worry. It's not totally sure whether you are only about to fall asleep—much too quickly—or that, slim chance though it may be, you're about to . . . die.

Like I said before, your brain does not joke around with mortal danger. So when in doubt, it sends a big jolt through your body to make sure you aren't about to kick the bucket. From your perspective, you feel like you've just had a major spasm—but you didn't—and you interpret it the same way as if you'd just fallen out of bed, because what just happened to you was complete relaxation—like in a fall—followed by a jolt running through your entire body—like at the end of a fall.

If I decided to write a whole book on the wonders of your brain and what it can do, I would still have to limit the scope of the book. But isn't it always like that when we get to the end of a chapter? And now it's time to discuss that absolutely critical subject: yawning.

58. Why Is Yawning Contagious?

To this day, we still don't understand why we yawn. We do know that yawning is instinctive. You only need to see a fetus yawning in its mother's womb to be convinced of that.

Is it about getting more oxygen to the brain? That's not it. Does it cool the brain? Maybe. Does yawning increase alertness? Stretch our muscles? Stimulate the production of a particular hormone? We don't know. We just don't know. No more than we know why, among all the vertebrates in creation, the giraffe is the only one that doesn't yawn. Fish yawn, as do birds, reptiles, dogs, cats, rabbits, bears, platypuses—platypuses, for god's sake!—monkeys, and humans. They all yawn. But not the giraffe. However, we do have a good explanation for why yawning is contagious and how the thing that makes yawning contagious is probably responsible for the "socialization" of humanity and is the basis of all civilization.

Empathy is certainly one of the best things about us humans as a social species. As it turns out, empathy is responsible for the contagiousness of yawning. Empathy is our capacity to place ourselves mentally in the position of another person. Apparently, there's a specific group of neurons responsible for this capacity; they're called *mirror neurons*.

Von Baer's Laws

Karl Ernst von Baer, a Russian physician, zoologist, and anthropologist, discovered some cool laws. They stipulate that during the gestation of a vertebrate, the general characteristics of the species are the first to develop, to then be replaced by the more specific characteristics. So fish and mammals are similar during the first stages of embryo development before they differentiate; then cows and primates are similar for a while until they differentiate, and so on. Yawning appears early in gestation, a sign that it is a very archaic mechanism that we share with numerous species.

Mirror neurons were discovered by accident in 1996—not all that long ago!—by a group of researchers led by Professor Giacomo Rizzolatti, an Italian neurophysiologist in Parma, Italy. After placing some electrodes in the brain of a monkey for experiments, Giacomo took a break to eat a sandwich. As he picked up his sandwich, he noticed the monkey's brain activity on the monitor indicated a movement. But Rizzolatti could clearly see that the monkey hadn't moved at all. She was just plonked there, vacantly gazing at the scientist, completely unaware that she had just written the first line of an important chapter in the history of humanity, the chapter on the biology of human social skills.

After observing this, Rizzolatti developed the following hypothesis, which he then confirmed: When the monkey sees someone perform a gesture that it recognizes and is able to do, part of its brain is activated the same way it would if the monkey itself were performing the gesture. The neurons involved in this phenomenon are the mirror neurons. When you perform an action, such as picking up a sandwich, this activates an entire area of your brain in the motor cortex—the part of the brain that controls your movements, hence the name. Interestingly, a portion of this area is activated, even though you're not performing the gesture, by simply watching someone else perform the gesture—and for that matter, even if you just think about doing it, it fires up that area of your brain.

This led Professor Vilayanur Subramanian Ramachandran, an Indian neuroscientist also known informally as "Rama"—and believe me I will be taking advantage of this—to ask a question: If, to the brain, performing an action and witnessing an action are the same, why isn't there any confusion? When I watch someone pick up an object, at no point do I think I am actually in the process of picking up an object. I have neither the feeling nor the perception of this action. So what's going on?

Well, I'll tell you. There's no confusion because our senses—in the following experiment, the skin sensors in particular—tell the brain that nothing is happening. No information comes from the arm about grabbing a sandwich. So the brain understands that it isn't really happening. As a result, we don't have any sensation of performing this action. Rama then tried an experiment to see whether this hypothesis was valid and, as you will see, every time our dear Professor Vilayanur S. Ramachandran tries an experiment, something amazing happens.

Rama completely anesthetized the arm of a patient and had him watch a person get pinched on the arm.

He hypothesized that the mirror neurons in the brain of the patient would be activated exactly as though his own arm were being pinched. At that moment, the brain would ask the areas that monitor touch whether the pinch should be perceived or not. In this case, though, since the arm is numb, the sensors wouldn't send any information to the brain to either confirm or deny the pinch. He thought, therefore, it might be possible that, being unsure, the brain would decide to activate perception of the pinch. And that is exactly what happened! The patient felt the pinch on his own arm, even though his arm was completely numb! The really incredible thing is, it's *only because* the arm was totally numb that the patient was able to feel the pinch. Rama, rather satisfied with his experiment, went on to test his theories on amputees.

When someone's arm or leg is amputated, it's quite common for them to continue to feel their lost limb. This is called phantom limb syndrome. Amputees can even give orders to their missing limb—to close their fist or bend their arm, for example—in what seems to be a completely normal way. And they really have—as an MRI will show—the sensation of these movements. But sometimes, the limb doesn't react the way it should—due to the fact that it's not there. This is when the phantom limb can become a real problem. Imagine a permanent itching sensation in an arm you no longer have, or that your fist is tightly clenched, day and night, without any relief and without your being able to do anything about it. Are you getting an inkling of the very real pain amputees may be suffering? Rama had an idea.

Suppose a person whose left arm has been amputated suffered from itching or muscle spasms in the missing arm. What happened if the amputee saw someone else scratch or massage their own left arm?

The amputee's discomfort eased, or even disappeared! Yep, the mirror neurons were activated, and there was no sensory information at all to draw on, because the limb was missing. Bolstered by his success, Rama decided to take on the "final boss" of phantom pains. Imagine your arm is missing and the hand on this arm is tightly clenched in a fist, all the time, without ever relaxing. Professor Ramachandran developed a contraption that's so complicated it puts to shame every Rube Goldberg ever created by *MacGyver*, *CrazyRussianHacker*,[150] or *Myth-Busters*. He took a shoe box, opened it up, put a mirror across the middle and cut two holes in one side, and covered one half. Done.

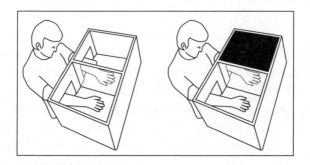

Ramachandran's therapy box

[150] YouTubers who specialize in homemade experiments.

Here's the idea: a patient places her existing hand through a hole in the side of the box so it can be seen in the mirror, and she places the stump of her wrist through the hole on the other side of the mirror—that side is closed on top. If she looks at it from the right angle, she can trick her brain into thinking the hand in the mirror is her missing hand. Once this has been accomplished, Rama asks her to clench both fists tightly, and then, after a few interminable seconds, he asks her to relax both hands at the same time. [...] A miracle takes place! The missing hand that had been clenched into a fist, perhaps for years, finally relaxes.

The pain disappears completely and immediately. But not necessarily permanently. So, to help his patients, Rama gives them his two-bit magic box to take home and use as needed. After a few weeks, the patients call Rama to tell him that the phantom limb has completely disappeared. Not just the pain ... no. The limb itself. They no longer have any sensation at all in the limb they no longer have.

So basically, the brain is again confronted with a contradiction. Sometimes, it believes the arm doesn't exist, but it hurts, and other times, the arm does exist—via the mirror—but it isn't transmitting any sensation. As a result, the brain goes into total denial and declares, "The arm does not exist. No further information about the arm will be processed. We are done with this farce."

I wouldn't be giving you the whole story if I failed to mention that sometimes the brain gets rid of only part of the limb. For some patients, their arm completely disappears except for their fingers. And since the brain isn't about to let fingers just go floating around in the air, it puts the fingers on the end of the shoulder. So these patients have the sensation that their fingers are attached to their shoulder and they can move them. That's right! They can do a Vulcan salute with nonexistent phantom fingers that stick out from their shoulder.

But that's not all mirror neurons do. They are also responsible for the preempathic state, a state in which we are more likely to be empathic and which promotes a social phenomenon called emotional contagion. As its name suggests, emotional contagion is a state during which an emotion can be quickly communicated and spread through a community, whether virtual or face-to-face. A study was conducted in 2012 on 700,000 unsuspecting Facebook users—all done in accordance with Facebook's general terms.[151] Here's the idea: Assemble two relatively similar groups of almost 350,000 users, and over the course of a few days, post articles and images that are more positive to one group and post more negative articles and images to the other group. During these few days, the users who received the positive posts seemed to be in a better mood. Those in the other group seemed to be in a worse mood. Thus it seems possible to transmit emotions, even to virtual communities made up of people behind screens.

Boris Cyrulnik, a French psychiatrist and psychoanalyst—and a specialist in the area of *resilience*—raised the question of the consequences of the significant amount of time humans are spending interacting via screens compared to less and less time spent interacting in real life.[152] His question: Can mirror neurons still work properly in the absence of direct human contact? An obvious example is written messages. Whether an instant message, text, email, tweet, or a comment on a social network, we've all had at least one experience where written expressions haven't conveyed—or conveyed very poorly—the author's intention. Without the right punctuation, intonation, and context, a message can be misunderstood: A bit of humor might be taken as a gratuitous attack. The fascinating thing is the ingenious way humans have resolved the problem. We've invented smileys and emoticons. So this sign :) means *I'm smiling*, o_O means *I'm confused*, and so on. We've managed to convert a facial expression into a short, simple, and immediate text.

Getting back to emotional contagion, you realize that once we understand the mechanisms, then we're ready to create a desired emotion in a population.

And mirror neurons are an awesome segue into neuromarketing and mass manipulation.

[151]   More information on this study at pnas.org/content/111/24/8788.full.
[152]   French radio program, *Histoire d'homme* on France Info, January 20, 2013.

Neuromarketing and Mass Manipulation

Neuromarketing is a field of research that straddles two disciplines:
neuroscience and marketing communications. There are two sides to
neuromarketing. Its primary goal is to better understand the cognitive
processes at work when a decision is made, such as buying something,
choosing a candidate at an election, etc., in order to recommend how an
advertiser should sell their product—or their candidate. And the other part of
neuromarketing is the search for better tools of persuasion.

Mirror neurons are an extremely powerful tool for communicators
because:

- They are found in all "normal" human beings—meaning, by the nasty
  normal word, people who are not completely devoid of sociability.
- They are unconscious, meaning an effective process works without
  the "customer" being aware of it.
- Their contagious nature expands the target market; once one
  person has been tuned in to an advertisement, he or she will be
  able to tune in others through their capacity for empathy.

The most eloquent example is the yogurt ad; I say *the* ad, because
almost all brands have put their knowledge of the neurosciences to good
use by designing ads that are basically the same: A woman eating from
a container of yogurt seems to derive a nearly sexual pleasure from it.
Viewers sitting in front of the TV will unconsciously associate yogurt
with the notion of an orgasm, and the ad may even activate the mirror
neurons in their brain linked with pleasure—without the viewer actually having the pleasure of the experience.

This way, a strong association is created in the mind between the
advertised yogurt and orgasm. At this point, you'd think the sale is practically in the bag, except that we are so used to and perhaps weary of the
multitude of ads like this.

Mass manipulation can work in the same way; television is still a very effective vehicle but is increasingly superseded by the internet. For example, consider the staggering number of alternative theories that are offered to explain the "Truth" about historical events by way of various conspiracies, be it the construction of the pyramids, the government cover-up about extraterrestrials on Earth, man having never walked on the moon, the delusional conspiracies about the death of JFK, oil sniffer aircraft, chemtrails, what really happened on 9/11, and so on. All of these theories pander to the ego and work by using a number of techniques well known to neurologists, psychologists, and sociologists.

First of all, they promise that you will learn something that not many other people know. Then, they provide a mountain of alleged evidence that leads you to say, "Well, it can't *all* be false," so you end up thinking, "Where there's smoke, there's fire, and there's a lot of smoke here!" But the most effective vehicle used by all of these theories, each more wildly elaborate than the last, is fear. Someone might express the fear that they are being followed and a price has been put on their head by every intelligence service in the world, or they may simply express the fear that the theory itself is supposed to arouse—suppose someone says extraterrestrials are already among us and have completed negotiations with the government to enslave us. Your mirror neurons will naturally, unconsciously, and completely and instinctively be activated. They make you feel, at least to some extent, the fear being conveyed. And that makes it much easier to accept the completely harebrained theory being offered.

Everywhere you see group behavior, whether within a flock of birds or a school of fish, mirror neurons are at work. You have probably already asked yourself: How can a group of peaceful demonstrators, over the course of a few minutes, turn into a bloodthirsty mob trashing everything in its path? Panic reactions, mass hysteria, or, in contrast, mass euphoria—particularly at sporting events—all of these are an effect of mirror neurons; it's called *herd behavior*. When a demonstration turns into a riot, we know very well that each individual demonstrator wasn't there to go on a rampage; but after a while, each individual no longer sees himself or herself as being with other individuals but as with a mob, and that's how each person ends up forgetting their own individuality. It's an amazing phenomenon, but it's explained simply enough: Each individual ends up behaving not like other people *are* behaving, but how he or she thinks other people *will* behave. These dynamics are also seen in the world of finance, where panic reactions can cause financial ruin for thousands of people in just a few seconds.

At the beginning of this section, I briefly mentioned that the processes to be discussed were probably the basis for civilization. How so? Don't you think that when an early human learned to manipulate fire—create it, keep it going, light a torch with it—there must have been another human there to copy those actions? It was only possible because of mirror neurons! The same applies to the first tools, the first clothing, the first drawings, the first buildings, the first agricultural knowledge, and so on. Without mirror neurons, it's not possible to become socialized or to transmit an emotion through empathy. Without mirror neurons, there would be no art and probably no language. It all began with the act of mentally and then physically copying an action. And it was probably even more difficult for left-handed people.

59. Left-Handers

The brain is made up of two hemispheres: the right hemisphere on the right and the left hemisphere on the left.[153] And once people became aware of this fact—the first time they opened up a head—and noticed that the two hemispheres were startlingly similar, it was normal for them to wonder if each half of the brain did more or less the same thing as the other half or if some parts of the brain were dedicated, on one side or the other, to specific functions. They wanted to know whether the brain was *lateralized*.

Paul Broca, a French physician and anatomist, discovered in 1865 that when there were lesions on the left frontal lobe of the brain, it had much more effect on speech than when similar lesions were present on the right frontal lobe. He concluded that the speech center is indeed

[153]   There must be some mnemonic device for remembering how it goes.

located in the left hemisphere of the brain, in the frontal lobe, in an area that has since become known as *Broca's area*. Similarly, in 1873, Carl Wernicke,[154] a German psychiatrist and neurologist, discovered that lesions in a different and larger area in the left hemisphere cause problems with understanding spoken and written language but don't affect one's ability to articulate sounds with their mouth. Thus, affected patients speak with a fluency that is as natural as it is incomprehensible. The area is called—surprise, surprise—*Wernicke's area*.

From there, it was just a matter of time before people started wondering whether there is ultimately a correspondence between each area of the brain and a specific function. Or, conversely, does each function—motor function, sight, smell, hearing, memory, etc.—correspond to an exact area in the brain? With the advances in medical imaging—and if I were talking about Nikola Tesla in this book, I would certainly have something to say here—scientists quickly realized that yes, some functions are very precisely localized in the brain, and moreover, it's largely asymmetrical, especially when it comes to intellectual functions—processing numbers, perceiving faces, and so on.

To understand the laterality of the brain, you need to know the difference between two things: the body's motor functions—its movements—and the sensory receptors involved, versus anything to do with thought and intellectual functions. Basically, when it comes to the muscles, organs, senses (everything to do with the body's "physical" functions), the *motor pathways* and *sensory pathways*—the pathways that connect nerves to the brain—cross from one side to the other at the brain stem, which is under the skull at the nape of the neck. The result is that a sensation or motion on the right side of the body is processed by the left side of the brain and vice versa. When it comes to the brain, you could say that the "physical" part of the body is completely lateralized. For the other functions, though, there is no such complete laterality. If we take the example of language, yes, it is strongly lateralized on the left, as we just saw above, but language isn't all on the left; the right hemisphere of the brain contributes, too. It controls language tone and word comprehension.

[154] Pronounced "vair-nick-ay."

There are some very old and very persistent theories that some people are more "left-brained" or "right-brained," and folks are in the habit of saying that an artist is more right-brained, whereas a good old-fashioned Cartesian accountant would be more left-brained. Sometimes, they even reverse the formula and say that a right-brained person is more artistic and a left-brained one is more rational and Cartesian. Either is absolutely false.[155] We are actually lucky not to be defined in this way.

Yes, things do work a bit like that, to the extent that someone who is particularly rational will be likely to work their left hemisphere more, just as someone who is especially creative will work their right hemisphere more. But remember what we discussed at the very beginning of this book: correlation and causality. And even if one imagines an artist to be more "right-brained," it doesn't mean that their brain was predestined to be lateralized to the right side (and thus it could only produce an artist).

And then there's something we don't hear enough about: the *corpus callosum*. This is the part of the brain that connects the two hemispheres. All of the wiring between the two hemispheres—which is essential—is in this corpus callosum. It connects the four lobes of each hemisphere two by two: moving from the front of the skull toward the back, you have two *frontal lobes*, two *temporal lobes*, two *parietal lobes*, and two *occipital lobes*. Lesions in the corpus callosum lead to coordination problems, whether physical or mental. For example, a *left-side ideomotor apraxia* prevents a person from carrying out a symbolic gesture using the left side of the body, such as a military salute. Another example, *left-side unilateral alexia* prevents the person from naming objects in the left side of their field of vision.

So are the areas reversed depending on whether you are right- or left-handed? Are the brains of left-handers a sort of mirror of the brains of right-handers? Or do we all use the same zones in the same hemispheres? And what difference does that make for left-handers? OK, first of all, no, the brains aren't reversed. Left- and right-handers normally have the same Broca and Wernicke areas in the left hemisphere. But

[155] So it's not so much right-brained as it is "wrong-brained" . . . Am I right? Huh? Anyone?

left-handers are less strongly lateralized, meaning they use both sides more, and that is very easily explained.

Left-handers represent less than 15 percent of the human population—BTW, in some animal populations, such as horses, the majority are left-handers, er, left-hoofers—which means that left-handers literally live in a right-handed world, built for right-handers. If you are a right-handed reader, you probably think I am exaggerating, so let me give you some trivial examples, so trivial that no one ever thinks about them—not even left-handers . . . most of the time. Imagine the start of a typical day for a left-handed student. After buttoning up his shirt like a right-hander—because to do it like a left-hander is almost impossible—and leaving home to go to classes, he heads down the stairs, like he does every day, to the metro station. To get to the train, he uses his handy-dandy transit pass, a smart card. All he has to do is wave it breezily over the reader on the turnstile . . . on the right. After this contortion, to which he pays almost no attention, he steps onto a train, but before the doors close, he notices the time on the platform clock. Ugh! He's late. Now, he sees his wristwatch shows a different time. After asking a fellow commuter for the correct time, he's now sure his watch is slow. He then puts his right arm out in front of him to adjust his watch with his left hand, by turning the knob of his watch, which is on . . . the right—he sticks his arm out, then has to bend it back in at the elbow so his left hand can get at the button or knob; so now his left arm is sticking forward, too, also bent at the elbow, struggling with his left hand to adjust a watch that's designed for the opposite of the way his hand is designed to achieve that operation. After a few seconds of this bizzaro version of the Funky Chicken, he takes off his watch and adjusts it with his right hand, just like he did at home when he bought it. It's just easier this way. During the trip, he puts in his earphones, connected to his smartphone, same as lots of other people, and he listens to his music. What if he doesn't want to listen to the song that comes on? No problem, he just has to press the button on his earphones twice to go to the next song. He does this, with his left hand, on the button on his earphones, which is on . . . the right.

Arriving—late—for his class, he quickly takes a seat in the lecture hall and folds down the desk attached to his chair . . . on his right. He takes out his spiral notebook and starts to take notes. His left wrist rests on the spirals and as he writes, his left hand covers up what he is writing. He completely ignores this; it's been like that since elementary school.

Back then, his wrist usually rested on the still-wet ink, smearing it across the page. So he learned to twist his wrist around into a hook to avoid making a mess. As a result, his handwriting is as elegant today as it was back in elementary school.

I could go on and on—being a left-hander, I find it quite entertaining . . . not. Let's talk about can openers, scissors, numeric keypads on the right side of keyboards, left-click on the mouse, pens and pencils on which the brand or advertisement appears right-side-up for right-handers only, the camera shutter button that's always on the right, the inside pocket of the jacket that's always on the left to be reached with the right hand, military actions demanding right-handers and left-handers move in the same way—that is, like right-handers, and so many more things ad nausea. But I think you get the idea. Left-handers live in a right-handed world. And ultimately, that's an advantage for left-handers. In fact, apart from having to learn to do a slew of things as a left-hander—write, walk, jump, run, swim, etc.—left-handers also have to learn to do myriad other things like right-handers do. This means that for anything to do with physical coordination, left-handers make *both* sides of their brain work more often than right-handers do. And that, my dear friends, is the reason why left-handers are less lateralized: They use both sides of their brain more equally. It is also the reason left-handers generally have a somewhat thicker corpus callosum.

Does that mean left-handers are more intelligent? Stronger? More agile? No, because being predisposed to use your brain more doesn't mean that you do it—for example, my legs predispose me to run, but that doesn't make me Usain Bolt[156]—and also because the brain is more complicated than that.

[156] You know, 100 meters in 9.58 seconds; I shouldn't even need to say it.

Though what makes someone intelligent is somewhat of a mystery even today, we know it has to do with the environment, family stability, education, and a multitude of parameters, each less important than the previous; however, it is entirely possible to have a very difficult childhood and still be intellectually brilliant.[157]

So, now we know that left-handers must continually adapt to a right-handed world. There's a word in that sentence that should make you look twice: *adapt*. Why the hell haven't the left-handers adapted into right-handers over the centuries and become extinct? Why aren't the right-handers, who are much better adapted to this world, the only ones still around? First of all, we need to decide whether this question is actually valid. After all, who can say for sure that left-handers haven't been in the process of dying out all along?

Well, actually, anthropologists and paleoanthropologists. Yes. They can. They've been able to figure out from studying prehistory, by examining tools and cave paintings left by our ancestors, that left-handers were not any more common five thousand, ten thousand, or fifteen thousand years ago than they are today. But where do left-handers come from? And why have their levels remained so consistent? Several hypotheses have been proposed on the origin of left-handedness.

Let's start with some questions. Does it have a cultural origin? No. No culture has ever to our knowledge intentionally privileged left-handedness, and generally the opposite is true. You just have to look at certain words to realize this. We have: Rightness is truth, justice, perfection. That's right! Righteous! Right on! The words *adroit* (skillful, clever) and *maladroit* (clumsy, awkward) come from "on the right" (*à droit*) and "bad on the right" (*mal à droit*) in French, which has also given us *gauche* (left) for "clumsy or tasteless." In Latin, *dexter* means both "on the right" and "favorable," whereas *sinister* means both "on the left" and "unlucky" or *sinister* as in "harmful or evil." The Latin word *dexter* has survived in *ambidextrous* and *dexterity*. In English, *right* is opposed to *wrong*, whereas what is *left* is, depending on usage, what has gone or what remains—and who gets excited about leftovers? Cultures

[157] Read Boris Cyrulnik's work on resilience in relation to this.

have never favored left-handedness. In Japan, in the past, a husband could demand a divorce if his wife was left-handed; some Native American tribes bound a baby's left arm to ensure they would not grow up to be left-handed, and so on. As recently as sixty years ago in France, left-handed school students were forced to write with their right hand, creating generations of "thwarted left-handers." No, the origin of left-handedness is not cultural and, if it were, left-handers would be a thing of the past by now.

Hereditary, then? Not that, either. Studies have shown that having left- or right-handed parents doesn't increase or diminish the chances of offspring being left- or right-handed.

Genetic, then? Yes and no . . . and maybe. It's a real mystery. Studies show that identical twins are just as likely to be one left-handed and one right-handed as any other two siblings. So, there's no gene for left-handedness. On the other hand,[158] other studies have shown that there are specific genes involved at the moment in gestation when an embryo changes from a completely round egg to a lateralized embryo with a well-defined left side and right side. It could be that these, or other genes, and possibly environmental conditions in the womb— position of the embryo, temperature, etc.—are responsible for the fact that a fetus in its mother's womb is already right- or left-handed. And handedness doesn't often change after birth.

But even if we still don't really understand the origin of left-handedness, is there an evolutionary reason why its incidence in the population has barely changed over the last five hundred thousand years? Think about this: Being left-handed in a right-handed world is, in a completely counterintuitive way, an enormous competitive advantage.

Consider any sport in which you face an opponent. If you're right-handed you will, about 85 percent of the time, find yourself opposite a right-hander. You get used to competing against right-handers. In tennis, for example, you know that if you send the ball to your opponent's left side, you'll receive a backhand that's less powerful than their forehand. When you encounter a left-hander, you use the serve you're in the

[158]   Right or left? Ha! Ahem, sorry.

habit of using and BAM!, you get a strong forehand to your own left side. If you are left-handed, the same thing can happen to you; since you're used to facing right-handers, you can be caught off-guard by left-handers just like any other player. The fundamental difference is that, because you are left-handed, you *always* find yourself playing someone whom you will surprise. And it's not just random chance that, among champions in tennis, Ping-Pong, fencing, boxing, baseball (considering only the pitcher and batter and not the rest of the team), soccer (considering only one-on-one situations, essentially dribbling contests, between two players), and so on, you find a higher percentage of left-handers than in the general population.[159]

On the other hand, in sports for which you compete alone—swimming, golf, track and field, etc.—the percentage of left-handers is the same as in the rest of the population, which is to say between 8 and 15 percent. It is also worth noting that in the sports for which different equipment, such as golf clubs, is needed for right- and left-handers, left-handers are more likely to use right-handed equipment, because that's most likely what was available when they started out.

A reader who is an evolution nerd will usually raise an objection at this point: If being left-handed represents such an advantage, the rate of left-handers should have naturally increased to 50 percent of the population—at above 50 percent, being right-handed would become an advantage, so we'd end up with an equilibrium of 50 percent. That's true, but being left-handed is not strictly an advantage. It is indeed a *competitive advantage*, but it's also a *collaborative disadvantage*. That's because people—the species, not the magazine—are not only a competitive species, they are also, as we saw in the previous section,[160] a social species in which individuals collaborate. And from a collaborative standpoint, left-handers are handicapped. When a right-hander makes a tool for cutting stone, it is a right-handed tool. The left-hander who

---

[159]   In fencing, the percentage of left-handers at the highest levels is almost 50 percent. Once this point is reached, being left-handed no longer offers any advantage because it affects half the population—it is thus in a state of equilibrium.

[160]   See previous section, obviously.

picks up this tool will be more *maladroit* using it. That's also part of evolution: sharing tools with your neighbor to share your knapping[161] skills, agricultural techniques, house-building methods, etc. If the world we live in were *only* collaborative, left-handers would have disappeared a long time ago, because it is a major disadvantage for collaboration.

As a result, we might conclude that if there has been a stable ratio between left-handers and right-handers for the last half a million years, maybe that means that people—the species, not the magazine—live in a way that's naturally both collaborative and competitive, and maybe in these proportions: 90 percent collaboration and 10 percent competition. Perhaps that's the natural balance that allows the human species to flourish.

60. A Conclusion About Life?

Absolutely no conclusion about life. I'm sure you can appreciate that we are a long way from being done with the subject, and if we want to know whether we can reasonably assume we are alone in the universe, we need to start by figuring out what exactly we are looking for when we talk about life. We have just scratched the surface of the subject of life *as we know it*, and there's a lot further to dig down before we can come to an informed opinion.

We haven't said anything yet about amino acids, proteins, DNA, RNA, living organisms that are so resistant to everything and anything that they are almost invincible—yes, tardigrades, I am looking at you. But the book must move on and we are only a few dozen pages from the end now. So the next chapters on life will have to be chapters of a whole other book. Except maybe one small last one in the next section. Thank you very much.

---

[161] Knapping (stone cutting), not napping. For left-handers and right-handers *napping* skills are shared with equal ease.

THERMODYNAMICS

Warmth is life's greatest gift.

Baloney! Hmmm, maybe that transition was too abrupt. The subheading for this chapter was supposed to make you think we were going continue talking about life. Did it fool you? We've departed the huggy-feely domain of life and love and are now returning to machines, engines, and gases that heat up, right? In one sense, yes, but you'll come to see that the two aren't mutually exclusive. The conclusion might even surprise you.

61. So, What Is it? Do Tell.

Thermodynamics is sufficiently foreign to you folks—well, most of you, probably—that I should spend a little time explaining what it's all about. To some extent, thermodynamics is the science of heat exchange, but that's not all. It's also the science of large-scale systems, but that's not all. The science of industry, yes, but that's not all. And it all began twenty-five hundred years ago when the Greek philosophers became interested in the nature of matter. Empedocles broke the world down into four elements: air, earth, water, and fire. And Aristotle came up with four fundamental characteristics—hot, cold, dry, and wet—to allow original matter to form these four elements. We're interested in hot and cold.

Back in antiquity, philosophers confused heat and temperature, and you can't really blame them because it's still easy to confuse the two, even today. Specifically, they confused heat with sensation of hot. But although these are easily mistaken, rest assured, it's also easy to tell them apart.[162]

---

[162] We covered that on page 47.

62. Is the Cake Pan Hotter Than the Cake?

Simple, homemade experiment: You bake a cake in a cake pan. When you take it out of the oven, you know it would be extremely unfun to touch the cake pan with your fingers. That's because it's been baking for an hour at 350°F—touching it would lead to unbearable searing pain. However, touching the cake to see if it's done, or fluffy enough, or who knows what, that's just fine. It's obvious to everyone in the world that the cake pan is way hotter than the cake itself. Or is it? After an hour in an oven that was preheated to 350°F, the cake and the pan have reached thermal equilibrium in the oven. They're both at 350°F. They are as "hot" as each other. Technically, their temperatures are the same—we can certainly measure that easily enough to confirm that. OK, but as I mentioned a moment ago, you can touch the cake, but you can't stand to touch the pan. What gives?

Another just-as-simple experiment: At your house, in the living room, in the office, in the sun on the deck, wherever works for you, put a book and a metal hard drive—you must have one lying around somewhere— next to each other. If you leave them in the sun long enough—the famous "sufficiently long time"—you know for sure the hard drive will be hotter. If you leave them in the shade in your living room, or on the coffee table, the hard drive will be colder than the book—common sense, it's what the thermoceptors in your skin will tell you if you touch them.[163] However, in both of these situations, after waiting long enough, thermal equilibrium occurs. Think about it—the book and the hard drive are at the same temperature. Go get a thermometer and measure the temperature and you'll see. How can your senses be mistaken about this? It's actually *because* it's your senses and therefore your perception. When it comes to temperature, your senses aren't for measuring the temperature of objects. They're for warning you that you're in danger of getting burned.

That being the case, what makes the cake pan feel so much hotter than the cake? It has to do with the material the pan is made of. Let's consider metal: It's a great thermal conductor. So, when it's hot, it "dissipates"

---

[163]   Thermoceptors are your heat sensors (see page 67 in chapter 14, on the senses).

a lot more heat over a given time than cake does—that's exactly why cake pans and pie pans are made of metal. Similarly, when you touch a metal pan that's colder than your hand, more heat will be dissipated from your hand to the metal object, so it seems more "cold," even though it's the same temperature as a piece of wood or plastic that's sitting right beside it. *Temperature* is a local measure of the agitation of the molecules in the material. And *heat* is specifically the transfer of energy between particles.

63. The First Steam Engine

Way before the Industrial Revolution and widespread use of steam engines, the first steam engine—that we know of—wasn't a big deal. But what it did was revolutionary. The *aeolipile* (also spelled *aeolipyle* or *eolipile*),[164] aka Hero's engine, was designed by Hero of Alexandria a little less than two thousand years ago. The machine was a water-filled sphere mounted on a horizontal axle, like a chicken on a rotisserie spit. The sphere had two open-ended bent tubes sticking out of it. They put the whole gizmo over a fire. What happened next might seem pretty basic to us, but for that time period, the experiment bordered on miraculous. The water heated up and produced steam that escaped through the bent tubes and that, of course, made the sphere spin around its axle. It may not sound particularly impressive—personally, I think you should be impressed . . . with yourself, because you realize, and Hero didn't, that the heat changed the state of the water, turning it into a gas,[165] which increased the pressure and forced the air and steam out of the tubes and—action, reaction[166]—pushed the tube, which made the ball spin.[167] See, you know all that!! Ha!! What a great book this is! You should give a copy to your relatives—but Hero of Alexandria was in the process of proving that heat can be transformed into work. Specifically, the heat from the fire changed the state of the water, and this change of state produced work.

---

164   Literally "Aeolus's ball," Aeolus being the Greek god of the winds.
165   Gas law (see page 184).
166   Newton's law of reciprocal actions (see page 183).
167   Converting linear motion into angular motion (see pages 195 to 196).

Work

We already had a detailed discussion about what work is (see page 189), but it can't hurt to kick it up a notch. The work of a force is the energy provided by the force during a displacement. This concept is going to be particularly important in thermodynamics when it comes to machines. Machines that produce work are called *motors* or *engines*, so it's easy to understand that work is directly associated with the concept of power, which is simply the measure of work per unit of time. It will be essential for being able to compare the efficiency of various engines to determine which engine is most efficient.

64. Good Old Science Dude: Francis Bacon

In the seventeenth century—I know, I know, we just made a freakin' huge jump in time—Francis Bacon was an English philosopher, who as early as the end of the sixteenth century—we're talking 1591, folks— wanted to improve the sciences ever since his time in Trinity College— yes, again!—where no one could ignore his obvious geniusness. And he did, indeed, play a large part in overhauling the sciences and founding what we call *modern science*; in fact, many historians call him, rather than Galileo, the father of modern science and the scientific method. Yeah, that big of a deal!

He planned to author a monumental work titled *Instauratio Magna Scientiarum* (*Great Instauration*)—whatever that means . . . actually I looked it up, it means the "action of restoring or renewing something"— with six parts to achieve this goal. He succeeded in writing only the first two of the six, but they alone were enough to open the way for a new mode of scientific thinking. The first part, *De Dignitate et Augmentis Scientiarum*,[168] discusses the classification of all the known sciences, their limits and their shortcomings. It definitely wasn't his most interesting work, but he did a nice job with the second part, *De Verulamio Novum*

---

[168]   *The Dignity and Increase of the Sciences*, 1605.

Organum Scientiarum,[169] or more commonly—and simply—known as *Novum Organum*.

In the second part, Bacon developed a new scientific method. It required that scientists free themselves of their prejudices and acquire, through experimentation, enough data to use inductive reasoning to infer the causes of a phenomenon and how they make the phenomenon happen. Thus he is one of the first postantiquity *empiricists*—ancient empiricism inspired very little Renaissance empiricism, other than the notable exception of Alhazen.[170]

Bacon divided his *Novum Organum* into books. In the first book, he pointed out that the sages in antiquity and the contemporary philosophers were not using any methodology at all, and this explained the lack of progress made so far. Then, in the second book, he elaborated on what he felt was an indispensable method and proceeded to lay, stone by stone, the foundations of what would become modern science.

In the first book, Francis Bacon maintained that experimentation was the most effective tool for understanding and mastering nature. But first researchers had to rid themselves of their *idols*, meaning their preconceived ideas, prejudices, and illusions about nature. There were four different kinds of idols. Today, we call them Bacon's Four Idols of the Mind:

Idola tribus: Idols of the tribe, shared by all humans, because our senses can be mistaken. These idols are the result of confusing what we perceive (subjective knowledge) and what actually is (objective knowledge).

Idola specus: Idols of the cave are related to each individual, created by one's education, beliefs, likes and dislikes, and habits, whether good or bad. Each individual sees the world through their own prism, and his allusion to Plato's *Allegory of the Cave* nails it.

---

[169] *The Advancement of Scientific Learning*, 1620. The Latin title is an allusion to Aristotle's *Organon*.

[170] A very notable exception indeed (see pages 35 to 37)!

Empiricism

Empiricism, in contrast to the rationalism of Descartes, Leibniz, and Kant, was based on the idea that experimentation is the source of knowledge. According to Francis Bacon (of course), John Locke, George Berkeley (alumnus of Trinity College, but the one in Dublin), and even David Hume, as well as other empiricists, knowledge was also the result of social experience. They believed that knowledge built up by many contributors, that new ideas were formed by combining existing ideas that came from different individuals. Empiricism was in strong and striking opposition with Aristotle's idea that experimentation was unnecessary—even useless—because, he believed, the answers were already around us, all he had to do was apply his deductive logic to understand them. It's interesting to note that when Bacon was only sixteen years old, he wrote a book that completely rejected Aristotelian philosophy. Dismissing his professors and instructors as "schoolmen,"[171] he said that a large number of them have sharp wits and an "abundance of leisure," but they have read only a few authors. It's as if their minds are imprisoned by the limitations of that small number of authors, he observed— as if they were confined to monastery cells, "knowing little history, either of nature or time"—and above all, he chided, their dictator is Aristotle.[172]

Idola fori: The idols of marketplace, or simply idols of public places, are idols of language and personal interpretations that lead to differences of opinion, misunderstandings, and difficulties in understanding each other.

Idola theatri: Idols of the theater, of the stage, come from tradition and authority, when the thoughts of ancient or famous authors— like Aristotle, infamously—are overvalued, closing the mind to new ideas.

---

[171]   Scholasticism was a teaching from the Middle Ages that reconciled Christian theology with ancient Greek philosophy, primarily Aristotelian.

[172]   *The Advancement of Learning* (1605).

Once these idols have been chucked in the river, Bacon used the second book to roll out, step by step, his new method. His method was based on *inductive reasoning*. And this inductive reasoning is why Francis Bacon is in this book—in the section on thermodynamics—yes it is, and until just now, frankly, the reason wasn't all that clear, was it?

Inductive Reasoning

In logic, inductive reasoning—we'll call it induction for short—tries to generalize a particular fact from repetition. By the way, strictly speaking, this reasoning is flawed. For example, if my whole life long, I only ever see gray cats and I conclude that all cats are gray, this happens to be wrong. However, as far as inductive reasoning is concerned, I was right to think that. Thus inductive reasoning must always keep in mind that a law is never proven; it is logical reasoning based on probability. That means all it takes is one single exception to disprove the law. Bertrand Russell, one of the fathers of modern logic, gives a brilliant example of the limits of inductive reasoning in his story of the inductivist turkey as told by Alan Chalmers:

> The turkey found that, on his first morning at the turkey farm, he was fed at 9 a.m. Being a good inductivist turkey he did not jump to conclusions. He waited until he collected a large number of observations that he was fed at 9 a.m. and made these observations under a wide range of circumstances.... Each day he added another observation statement to his list. Finally he was satisfied that he had collected a number of observation statements to inductively infer that "I am always fed at 9 a.m." However, on the morning of Christmas Eve he was not fed but instead had his throat cut.[173]

And this has to do with thermodynamics how? OK, I'm getting there. Inductive reasoning is based on probabilistic data, and that's the connection with thermodynamics, because thermodynamics is the science of large-scale systems. We're getting there, patience, patience. But first, it's time to hear about some good stuff from Bacon's second book:

[173]   Alan Chalmers, *What Is This Thing Called Science* (Milton Keynes, UK: Open University Press, 1982), p. 14.

specifically, his idea for three "tables of comparative instances" as a method for organizing and interpreting data.

The Table of Presence: When we study a phenomenon, we fill out this first table by noting all the situations where the phenomenon is naturally observed. For example, regarding heat, we could write sunlight, fire, rubbing our hands together, etc.

The Table of Absence: In the second table, just the opposite, we write down the situations, similar to the other tables, where the phenomenon was not observed. Continuing with the example of heat, we'd write moonlight, which doesn't warm things.

The Table of Degrees: In the third and last table, the list of situations where the phenomenon was observed is organized by degree of intensity. For example, sunlight heats less than rubbing your hands together, which heats far less than a fire does.

By bringing together the information in these three tables, we can infer, or generalize. Using the first table to establish the cause of the phenomenon, using the second table to find what's missing, along with the third table, we can observe what changes when the intensity changes. From there, a hypothesis can be generated and experimentation can begin to prove the hypothesis. This method is Francis Bacon's *new instrument*. Using this method, he concludes—rather, infers—that heat is related to motion:

> What we have said with regard to motion must be thus understood, when taken as the genus of heat: it must not be thought that heat generates motion, or motion heat (though in some respects this be true), but that the very essence of heat, or the substantial self of heat, is motion and nothing else.[174]

---

[174]   Francis Bacon, *Novum Organum* (1620).

65. Sadi Carnot, Father of Thermodynamics

Nicolas Léonard Sadi Carnot, known as Sadi Carnot, was a French physicist and engineer at the dawn of the nineteenth century, a graduate of l'École Polytechnique (the Polytechnical School). He wrote only one book in his entire life. His *Reflections on the Motive Power of Fire and on Machines Fitted to Develop That Power* came out in 1824, and even though he was only twenty-seven years old, it was his life's work and the basis of a new scientific discipline, thermodynamics.[175]

It's one of those rare cases in which a science is developed in a theoretical way around the desire for practical, industrial uses—that is, converting fire's energy into mechanical power. Of course, his book had its imperfections, as any groundbreaking work will. Specifically, Carnot included theories about heat that were popular at that time, specifically the *caloric theory*.

A bit like Newton in his treatment of optics, Carnot wasn't as interested in the nature of light as he was in the laws that governed its behavior. In his work, he focused on heat as an energy source, as a *heat engine*, without really worrying about its nature. A fluid? Not a fluid? Who cares? As long as it works. Heat was a real phenomenon that could be measured—the thermometer was invented in the seventeenth century and, yes, we still manage to confuse heat and temperature—so, it made sense to be curious about the laws that governed heat's occurrence, movement, and disappearance.

Realizing—he was one of the first—that even the best steam engines had a negligible output compared to the forces of nature that heat could produce—wind, ocean currents, hurricanes, etc.—Carnot focused on the idea that heat was the primary cause of these natural phenomena. With his background and education, Carnot was one of those rare individuals able to combine philosophy, steam engine design, and meteorology. The first part of his book concluded that anywhere there is a difference in temperature, you have what you need to generate

---

[175] Actually, the term *thermodynamic* was invented later by William Thomson . . . you know him . . . Lord Kelvin!

Caloric and Phlogiston

At the end of the seventeenth century, *phlogiston* theory tried to explain combustion by the existence of a fire element, a fluid present in any combustible body. This element was called phlogiston,[177] and better yet, substances that contained phlogiston were considered to be *phlogisticated*—what a word!—as things burned and lost their phlogiston, they became *dephlogisticated*. We can still see that ol' Aristotelian desire to find the element fire in matter. However, the observed loss of mass after combustion seemed to confirm the departure of the phlogiston.

Many experiments later, though, the shortcomings of the phlogiston theory became apparent: particularly, the fact that magnesium actually *gains* mass upon burning rather than losing it. Some folks tried valiantly to explain this using elements with negative mass. The system finally crumbled when Lavoisier showed combustion required oxygen, and that laid the foundations for a new theory of combustion, the caloric theory. Though this theory had already been mentioned by the Scottish chemist Joseph Black, it was actually Antoine Lavoisier who introduced it, suggesting that heat is an imponderable and indestructible fluid called either *igneous fluid* or *caloric*—it's not *caloric fluid*, it's just *caloric*.

Lavoisier said: In general, we know that all bodies in nature are bathed in caloric, surrounded by it, saturated with it, and it fills any available space between their molecules[178]

motive power. This idea is the very heart of thermodynamics! From this conclusion, he also inferred that it's not possible to generate motive power, aka impelling power, without having a hot body and a cold body. This is actually a primitive rendition of the *second law of thermodynamics* (see page 182). And this is the angle he used to define the ideal engine in the second part of his book

[176]   This ridiculous name comes from φλόξ (*phlóx*), which means "flame" in Greek.

[177]   Lavoisier, *Traité élémentaire de chimie* (*Elementary Treatise on Chemistry*), 3rd. ed. (1780).

He envisioned an ideal machine, known ever since as the *Carnot heat engine*, which alternates between exchanging heat with a hot body and a cold body. The operation of this heat engine used an engine cycle that produced mechanical energy in the form of work. Here's how the cycle works. Imagine a closed cylinder—one end of the cylinder is closed by a piston—and the cylinder contains a gas we'll call the "working substance." Now, put a cold body and a warm body next to the cylinder, but not touching it yet. The engine we're talking about is supposed to be ideal, so we'll make the assumption that the piston can move without any friction. Put the hot body in contact with the cylinder, push on the piston to compress the working substance, and now we're ready to start.[178] The cycle has four steps.

Step 1: The substance expands, the pressure goes down, and as it expands it pushes the piston, which moves freely (note: when expanding, a substance cools), but in the cycle devised by Carnot the working substance takes heat from the hot body, so the temperature of the working substance doesn't change at all. This is called *isothermal expansion*—*iso-* is a prefix meaning "equal or same," used here because the temperature stays the same throughout the process . . . keep an eye out, you'll see *iso-* again.

Step 2: We move the hot body away from the cylinder, which means there's no more heat exchange between the working substance and the hot body—actually, we insulate the cylinder, here, and there's no heat exchange at all with the outside world. In this step, the substance continues to expand, and as it expands, it begins to cool. Because there's no thermal energy exchanged between the cylinder and the outside world during this expansion, it's called an *adiabatic expansion*.

Step 3: We place the cylinder in contact with the cold body. This is a compression step, so the pressure is going up—when pressure goes up, temperature goes up[179]—but in this cycle, the working substance gives heat to the cold body to keep the temperature the same as the pressure increases. That makes this an *isothermal compression*.

---

[178]   It's a cycle, so it doesn't really matter which step we use first.

[179]   Remember the compressions in the life of a star in chapter 24 (see page 96)?

Step 4: You get it . . . we now remove the cold body. Compression continues in this step. And because we've insulated the cylinder again, the working substance can no longer exchange heat with the outside world. You tell me what happens during this step—right! pressure goes up, temperature goes up—you get it. This step is an *adiabatic compression* step, and now, we—and the cylinder—are in position to start step 1 again.

I do realize all this seems rather complicated, but really, this notion of an ideal engine—getting work from a piston by transferring heat from hot bodies to cold bodies—opened the way for the whole field of thermodynamics. I want to point out that the Carnot engine uses a process with four steps that can be run backward to use the mechanical work of the piston to move heat from the cold body to the hot body. When you run it in reverse, it's called a *refrigeration cycle*.

If there were a loss of heat, aka wasted energy or dissipation, during any step, it wouldn't be a perfectly reversible cycle. A perfectly reversible cycle means the amount of work provided by a given heat transfer in one direction exactly matches the amount of heat transferred for the same work in the other direction. Since the cycle uses perfectly reversible steps—because it's ideal and there are no losses—the efficiency of Carnot's ideal engine is the highest efficiency you can get from an engine of this type.

Efficiency

Efficiency is exactly what you imagine it is. It's the relationship between the energy provided to an engine and the energy produced by it. A Carnot ideal engine has a high efficiency, which always depends on the temperatures of the hot body and cold body. Efficiency is always, always less than 100 percent—you can measure efficiency in percentage or a number value—a dimensionless number, without any units of measure—between 0 when efficiency is zip, and 1 when the engine is totally efficient—which happens . . . n e v e r. Therefore, real engines based on the Carnot engine have a lower efficiency—at best equal—to the theoretical efficiency of a Carnot engine.

Today, this theoretical cycle is very well known to physicists as the *Carnot cycle*.

When Carnot's book was published in 1824, it received a lukewarm reception from the scientific community—they just didn't get it. Important centers of excellence like the Institut de France and even his own school, l'École Polytechnique, totally missed the point. Only a rare few foresaw the far-reaching impact of this work. By and by, a certain William Thomson, our buddy Lord Kelvin, noticed his work. Working with Rudolf Clausius, a German physicist, they laid the groundwork for the field of thermodynamics using the *law of conservation of energy* as its foundation instead of the caloric theory. And the law of conservation of energy is none other than the *first law of thermodynamics*—after a long gestation, another branch of science was officially born!

66. The Three Laws of Thermodynamics

The discipline known as thermodynamics is based on a three-part foundation: the three laws of thermodynamics. But before we list them, we'd better define what is called a *thermodynamic system*—it's a kind of hard to simplify this, but here goes: A thermodynamic system is a portion of the universe that we've isolated by thought. What isn't in that part of the universe is called the external environment or surroundings (the outside world)—like I said, not so easy to simplify, right? Also, we say the system is *closed* if there's no exchange of matter with the outside world. And it's *isolated* if there is no exchange of energy—in the form of heat or work—with the outside world. In all the other cases, the system is deemed to be *open*.

First Law of Thermodynamics: Conservation of Energy

When a process takes place in a closed system, the change in energy of the system is equal to the amount of energy exchanged with the surroundings. It can be thermal energy—heat—and/or mechanical energy—work. This law tells you that if your closed system loses energy, that energy is gained, in one form or another, by the surroundings. This might seem blatantly obvious to you today, but at the time, it was revolutionary! It's the energy equivalent of the famous "nothing is lost, nothing is created" statement.

And this law gives us the opportunity to get into another very important fundamental concept in thermodynamics—number one is heat exchange— that is, understanding that a system that, on a macroscopic scale, seems to be at rest, at equilibrium actually contains myriad things (particles, molecules, etc.), which, at their scale, are not in any way shape or form at rest at all. Because thermodynamics is also the science of large-scale results from systems, it gives us insight, at a macroscopic scale, about exchanges taking place on an elemental level. It's no accident that thermodynamics, as a probabilistic science, is applied to highway traffic management, and that's just one example of how far-reaching thermodynamics really is.

Getting back to the first law of thermodynamics: *thermo*, as it's affectionately called, also allows us to describe, at a microscopic scale, the difference between work and heat. To say it simply, work is an orderly exchange of energy between the system and the surroundings, while heat is a similar exchange of energy, but a chaotic one. The concept of heat is intrinsically tied to the idea of agitation, or excitation— soon we're going to talk about Einstein's second 1905 article (see page 273), then you'll see what's so exciting about that.

Second Law of Thermodynamics: Entropy

We've already had a brief encounter with entropy (see page 201). Here's the law about it: Any process in a thermodynamic system is accompanied by an overall increase in entropy. That means the entropy in the system and entropy in the external environment. In other words: Any process in a thermodynamic system increases entropy. If you say it that way, it seems like no big deal. So let's have a closer look.

When a process is reversible, there's no increase in entropy—the change in entropy is zero. To illustrate these words of wisdom, let's do something simple: Please mix some hot water with some cold water. Wait a few minutes—time makes it possible to achieve the desired process— and you have lukewarm water. Bleh. There's no way to reverse this process without expending more energy than you used to get lukewarm water.

Bleh. When the two kinds of water were combined, entropy increased. A reversible process must be able to reverse what it did. This implies that the "disorder" doesn't increase during the process, because if it did increase in one direction, it would have to decrease in the opposite direction, and that's just not possible. So, the only possible conclusion: During a perfectly reversible process, the change in entropy is zero. In ideal cases, this works just fine. But in the real world, "real" processes are never ideal; there is always an accompanying increase in entropy.

There's No Such Thing as a Perfectly Reversible Process

In the real world, systems are way too complex for anything to be perfectly reversible. Even down to the molecular level in a liquid or a gas, molecules move pretty freely, and they certainly are never in the same place as when they started. Plus there's even infinitesimal friction between the molecules of a liquid or gas as well as between it and the solid container it's in, and so on; it all dissipates heat. And there are always chemical reactions going on in a thermodynamic system—a very small amount perhaps, but they are still taking place. That's why perfectly reversible processes are only an idealized mathematical goal.

Note: The entropy of a system can become less during a process, but the second law of thermodynamics is talking about overall entropy. This means that when the entropy of a system goes down, the entropy of the surroundings increases by at least that much. In the special case of an isolated system, where no matter, no heat, no nuthin' is exchanged between the system and the surroundings, random agitation on the microscopic scale will spontaneously tend toward *thermal equilibrium*, meaning homogenized agitation and, therefore, homogenized heat within the system. This homogenization is irreversible—going back to the experiment of mixing hot and cold water, put them together in a thermos and you have an isolated system—its entropy increases during the process. Thus entropy of this system is at its maximum when it gets to the point where the temperature is the same everywhere in the system. This is *thermal equilibrium*.

Third Law of Thermodynamics: The Nernst Law

I'm bringing up the third law of thermodynamics so we can say we've pretty much covered thermodynamics. But actually, this last principle really isn't required to do classical thermodynamics. It pertains only to a special case of thermodynamic systems: systems as their temperatures approach absolute zero (see page 48) and they're tending toward their ground state. The law was discovered in 1906 by Walther Hermann Nernst, a German chemist; it's also called *Nernst's theorem* or *Nernst's postulate*—it earned him a Nobel Prize in chemistry in 1920. This is what the law says: The entropy of a perfect crystal at absolute zero (0 Kelvin) is zero.

Let's look a little closer—but not too close—at what this means. Absolute zero, or 0 Kelvin (or just 0 K), is unattainable. It's supposed to be the temperature at which a system, at a macroscopic scale, no longer has any thermal energy, no more heat. This means that the particles in the system are all in their ground state, meaning their lowest-energy state. The particles become completely imperceptible and "motionless"—in the usual sense of the word *motionless*—and its entropy is zero. It is interesting that quantum mechanics requires there be a small amount of continuous motion, so it doesn't allow such a state. Be that as it may, using this law Nernst realized two important things: First, he realized he could give an absolute value to entropy; until then, only change in entropy was measured, so a given value of entropy was always relative, just like temperature was before the Kelvin scale was invented. Next, he realized that he could tell what would happen to a thermodynamic system when it approached absolute zero—and the truth of this law can be measured experimentally at extremely low temperatures. The record is on the order of one tenth of a nanodegree above 0 K—aka on the order of .1 trillionth of a Kelvin.

67. And Boltzmann?

Those who have studied thermodynamics are probably surprised that they haven't seen Boltzmann's name come up sooner. You see,

Boltzmann is as naturally associated with thermodynamics as Newton is with gravity and as Maxwell is with electromagnetism. But Ludwig Boltzmann, an Austrian physicist and philosopher, is as much of a special case as any of the other the illustrious figures I've mentioned in this book. As a matter of fact, Boltzmann was in line for the same destiny as Mendeleev, but unfortunately not everybody is cut out for that. When Mendeleev laid out his periodic table, the scientific community was still widely divided between the atomists and the non-atomists, but Mendeleev was made of steel, impervious to anything except scientific arguments, which his detractors didn't have—and for a good reason: He was right all along. Boltzmann was like a Mendeleev, except Boltzmann was sensitive and emotionally fragile—he was deeply affected by criticism. Sadly, this caused his death.

Boltzmann was interested in thermodynamics and, more specifically, in entropy. He saw a definite relationship between the way a thermodynamic system spontaneously organized itself and all the possible ways it could be arranged. While others may have noticed it earlier, it was indeed Boltzmann who made the connection between thermodynamics and statistics. It was important to him because Boltzmann was a confirmed atomist, and he was sure it would be possible to link the thermodynamic macroscopic behavior of a gas, with its multitude of atoms and molecules, to its behavior at a microscopic level.

Most of the scholars studying thermo at the time didn't need to know the underlying reality—at a microscopic scale—of a system in order to observe overall behavior and identify thermodynamic laws, no more than Newton needed to understand the nature of light to explain reflection. But Boltzmann built on Lavoisier's and Avogadro's work and proposed a new theory, known today as *statistical mechanics*, to accurately explain the macroscopic behavior of thermodynamics using the microscopic properties of the systems. In 1877, he established a relationship between microscopic and macroscopic properties with a formula that's so simple and elegant that I can't resist writing it:

$$S = k \cdot \ln W$$

This equation is so marvelously brilliant that it is etched on Ludwig Boltzmann's tombstone in the Vienna Central Cemetery, the *Zentralfriedhof*—in the form that was used at the time: S = k. log W.

What does it say? Imagine that you have a liter of a given gas. Thanks to Avogadro, we've known since 1811 that this liter of gas contains a given number of molecules (see page 14). At standard temperature and pressure that's about 30 sextillion molecules. Each molecule moves pretty freely, and the number of options for the ways they can arrange themselves is majorly colossal!

By making a statistical estimate of the possible arrangements—that's the W in the formula; it's a very big number—which he called the *microstates* of the system, Boltzmann determined that there was a mathematical relationship between the entropy of a system—S in the formula—and this enormous statistical number. The equation says entropy, S, is proportional—by a constant, k, called *Boltzmann's constant*—to the natural log of that huge number W. In addition to establishing a relationship between the microscopic and macroscopic, this formula makes it possible to use entropy to find missing information about a system, because the amount of entropy tells you the number of possible arrangements of molecules in the system. So if we refer back to the third law of thermodynamics, zero entropy means—because k is a non-zero constant—that ln W is also zero, which means—I'm so sorry about the math, but it's almost done—that W = 1. When entropy is zero, it means that there is only one possible arrangement for the parts of the system.

Here, we again see the idea—always unobtainable, remember—of the third law of thermodynamics that at absolute zero, particles are imperceptible and motionless. Boltzmann was criticized by many detractors who scoffed at his discovery. This plunged him into major depression and drove him to try to kill himself; his second attempt was fatal. He did not live to see the triumph of his ideas in thermodynamics as well as in fluid mechanics. It came about particularly through Planck's work on black bodies (see page 49) and Einstein's work on Brownian movement.

A Tiny Little Bit of Statistics

Imagine a deck of cards, a regular deck with 52 cards plus 2 jokers. If you shuffle the deck of cards, how many possible combinations can you make? The math is easy enough: the first card from the deck can be any one of the 54 cards, so there are 54 possibilities for the first card chosen. Then the second card can be any of the remaining 53 cards, so for each of the 54 first possibilities for what the first card would be, there are 53 possibilities for the second card; this makes 54 × 53 = 2,862 . You continue the same way until there's only 1 card left. The total number of possibilities is 54 × 53 × 52 × 51 × . . . × 5 × 4 × 3 × 2 × 1.

There's a mathematical notation—a symbol—called factorial, designated by an exclamation point. It lets you write that long multiplication in a simpler and more concise way. Factorial means that you multiply the number by all the whole numbers that are less than it, all the way down to 1.

So for example: 5! = 5 × 4 × 3 × 2 × 1 = 120

The total number of combinations for a 54-card deck of cards is

54! = 230,843,697,339,241,380,472,092,742,683,027,581,083,278,564,571,8 07,941,132,288,000,000,000,000

This entirely grotesque number represents the possible arrangements of only 54 cards! Now imagine how many different ways the 30 sextillion molecules in one liter of gas can be arranged.

68. Einstein in 1905: The Second Article

Well, here we are. We've made it. It's 1905, Einstein's miraculous year. He sent in his first article in March—published in June—on the photo-electric effect. And a few weeks later, in May, he sent a second article to *Annalen der Physik*,[180] and it was published in July. The second article

[180]   "On the Motion of Small Particles Suspended in a Stationary Liquid, as Required by the Molecular Kinetic Theory of Heat," *Annalen der Physik*, vol. 322, no. 8 (July 18, 1905).

was about Brownian motion. About what?? Take a glass filled with hot water and sprinkle in a few grains of pollen. You'll see the pollen dancing around in the water in a surprising way: abrupt changes in direction, sudden stops, sudden starts, etc. This erratic motion is called *Brownian motion*.

Einstein was convinced that atoms were real—the scientific community still considered atoms as a mathematical tool, more useful for equations than a physical reality—and he therefore theorized that hot water is made up of a multitude of very agitated molecules moving a little bit in every direction, and in the glass of water I mentioned above, these molecules frequently collide with the pollen. He presumed the pollen was moving because of these collisions. The biggest problem Einstein had was getting thermodynamics and classical mechanics to agree, because classical mechanics describes the motion of individual bodies while thermodynamics studies large-scale behaviors of systems.

He therefore used Boltzmann's work on the kinetic theory of gases—even though this work hadn't been accepted by everyone—to demonstrate his idea mathematically. He definitely gave a strong indication, an "almost proof," of the reality of the atom—Jean Perrin had the honor of providing the irrefutable proof of this reality. With this article, however, Einstein demonstrated there was complete consistency between Boltzmann's work and experimental observations.

Einstein, for the second time in the course of a few weeks, had just revolutionized the way scientists looked at nature. After taking on light and nearly demonstrating the existence of the atom, all he had left to do was completely blow up the basic framework of physics, that of space and time.

SPECIAL RELATIVITY

Despite what you've often heard, everything is not relative.

In previous chapters, particularly when looking at classical mechanics, we saw a long parade of illustrious scientists whose work vastly improved our understanding of the world we live in—the universe—as well as the rules that govern its behavior. I don't want to break up the party, but I have to tell you, everything they discovered is by and large false. Well . . . not really false . . . let's just say that everything they discovered is, in fact, very limited. All the rules, laws, and principles are valid only within a limited framework.

Trying to get outside that box has forced us to rethink everything we thought we were rather certain of about the universe. Lots of people have played a part in our new conception of the universe, but there are two figures who stand head and shoulders above the rest and whose genius has led to a definitive shift in paradigms. They are Giordano Bruno and Albert Einstein.

69. To Move or Not to Move? That Is the Question

Actually, moving or not moving is relativity. Do you remember Aristotle's proof that the earth is immobile (see page 160)? And do you also remember Giordano Bruno's proof that Aristotle had goofed once again (see pages 166 to 167)? Well, that's where relativity all started. Bruno did a very good job of showing that the state we think of as immobile is no different from a state of motion, as long as it is steady motion. From that point on, people talked about uniform rectilinear motion: movement in a straight line, without any variation in speed. Despite his extraordinary intuition, however, Bruno didn't demonstrate it or explain it in any kind of especially convincing manner. Same with his model of an infinite universe in which neither the earth nor the sun occupy a privileged position at its center. What Bruno said is that we are here, but we could just as easily have been anywhere else.

Soon would come Galileo and his very similar but more advanced experiment (see page 175) to reach the conclusion that would be the first stone paving the road to the special theory of relativity. Based on his experiment, Galileo concluded that if you don't feel any motion—which is to say acceleration—it is impossible, without a reference point, to figure out whether you are moving or not. This was a significant discovery! It demonstrated for the first time that the laws of physics are the same whether you are moving or not, and as a result, being in uniform rectilinear motion is exactly the same as not being in motion at all. Galileo even developed mathematic tools to work out what constitutes a valid "reference point," which we now call a *Galilean reference frame*, or an *inertial reference frame*.

Inertial Reference Frame

First of all, what is a *reference frame*? It is a system of coordinates that allows the position of any body—but usually the body being studied—to be defined in space and time. It has four parts, or four dimensions, three of which are spatial—typically length, width, and height—and one temporal, or time. This way, if I perform an experiment in a laboratory—suppose I drop an object on the ground, for example—I can consider the origin of this reference frame to be a corner of the room, and time zero is the instant I let go of the object. Officially, an inertial, or Galilean, reference frame is a reference frame in which the principle of inertia is valid, meaning in this reference frame, if a body is not subject to any force—or if the forces applied to it cancel each other out—it is either immobile in time or in uniform rectilinear motion.

The good thing about the principle of inertia is that once you have an inertial reference frame, any other reference frame that is not moving or is in uniform rectilinear motion with respect to the first is also an inertial reference frame. Going back to the example of Galileo's experiment with the boat, an observer on the riverbank with a watch is an inertial reference frame; Galileo onboard the boat is another one. All we ask of our reference frames is that they don't start changing speed or direction—and there must be no talk of acceleration. In any case, in Galileo's time, the very notion of acceleration didn't even exist yet. They were still bickering over the idea of impetus (see pages 153 to 155). So, Galileo developed mathematical transformations—geometric, in fact—to go from one Galilean reference frame to another while ensuring the inertial nature of the reference frame isn't lost.

As familiar as they may seem to a high school student today, these transformations were extraordinary. At the end of the seventeenth century, when Newton presents his laws of motion, we find Newton's laws are valid in reference frames related by Galilean transformations—this is known as Galilean invariance—which means Newton's laws hold true in any Galilean reference frame, and that is very cool.

While I'm still convinced Bruno was the first person to envision relativity, I have to admit that Galileo actually invented it right then and there. *Galilean relativity* amounts to saying that there's no absolute position, absolute speed, or absolute motion.

Galilean Transformations

First of all, and quite naturally, Galileo assumed that time is absolute. You can assign time zero to whenever you want, but a second is a second. That means that if two reference frames have a different time zero, the time discrepancy between the two remains constant. So if, at a given moment, reference frame R is 10 minutes ahead of reference frame R', R will always be 10 minutes ahead of R'.

Next, about the spatial dimensions of a reference frame, Galileo allows a reference frame to be rotated in any direction; on the condition, however, that the rotation stops once the reference frame is defined—a rotating reference frame is not Galilean.

Finally, a reference frame is permitted to move in rectilinear uniform motion, which means a constant movement at a constant speed can be applied to it.

Galilean relativity tells us that "the train pulls away from the station" is exactly the same thing as "the station pulls away from the train."

Now everything depends on the observer's point of view. For the first time, the observer—and thus the observation—is at the very heart of science. This is why Galileo is considered to be the father of modern science.

70. The Problem with Light

In 1687, in his *Principia Mathematica*, Newton expressed the principle of inertia mathematically. The statement of the principle of inertia is nothing less than the mathematical expression of Galilean relativity applied to the laws of motion. And for almost two centuries, Newton was untouchable, because his equations so precisely predicted every movement studied, from the motion of the planets to falling objects to the trajectory of a projectile. People had come to think that, more than just developing a theoretical model, Newton had found the underlying *truth* of physics.

As a result, Newton's theory of space and time was never questioned. To Newton, space was absolute, not in the sense that it has a point of origin—a point zero—but in the sense that space is like a backdrop. It is a rigid concept in which lengths are rigid, the same regardless of the reference frame. There's nothing shocking about that, after all; who would imagine that a meter for an observer sitting on a bench is not quite the same as a meter for an observer on a motorcycle? In the same way, according to Newton, time is absolute, not in the sense that it has an origin, but in the sense that a second is a second for everyone, whatever the reference frame of the observer. Time flows in a consistent way at the "speed" of one second per second. Eventually—after about two hundred years, that is—these very sensible ideas ended up causing some problems.

When Maxwell united optics, electricity, and magnetism, thus founding electromagnetism (see page 89), he suggested that light is a wavelike phenomenon, contrary to Newton's hypotheses (see page 42). But Newton's formulas had by then been in use for almost two hundred years, on a daily basis, with, as I've already mentioned, amazing accuracy, explaining the movements of everything you could see: the movement of the planets, falling objects, and the trajectory of a projectile. Newton's laws were so powerful that people had come to think, "That's it, we've found the *true formula*, we have a perfect knowledge of the laws that govern motion." Let me tell you, anyone daring to doubt the great Isaac Newton had better be armed to the teeth and quick on their feet.

The fearless Maxwell presented his equations, which were a continuation of Huygens's work on the nature of light, and in the end it didn't do much damage to Newton's theories—for the time being, anyway—as long as Maxwell could demonstrate that Newton's laws of optics were still valid even using a wavelike light. The problem was that back then, the only known waves were mechanical waves, which need a medium to propagate in—sound waves in air, ripples on the surface of the water, musical notes on guitar strings, and so on. So they naturally wanted to figure out what medium light propagates in, especially in outer space, which is supposed to be empty. The trouble all started with this medium that was supposed to propagate light—*luminiferousether* (or *luminiferous aether*), or simply ether.

71. The Ether

Just so you know, in Maxwell's time no one had ever even imagined a wave that propagated without a medium; the very definition of a wave was the disturbance of a medium. And on top of that, Maxwell's equations, the very ones used to deduce that light is a wave, worked fantastically well. It was natural to think: If light is indeed a wave, it should propagate in something.

Then came the questions: What is ether? What is it made of? Is it solid? Liquid? A gas? Something else as yet unknown? What is its mass? Does it move? If so, how? All these questions were the subject of numerous research papers at the end of the nineteenth century, each less informative than the last. There are only three things they managed to figure out about ether, and those were fairly obvious. As for the rest, nothing. Zip. Zero. Nada.

The first characteristic of ether is that it has to be rigid and stretchy at the same time. For example, when we speak, our voice doesn't carry dozens of miles. That's because sound is attenuated as it propagates through the air. But we know we can see light from the sun, and that's 93 million miles away, and even from stars that are several thousand light-years away! For the light to reach us, ether, in which light propagates, must barely attenuate it at all. So ether must be extremely rigid. Extremely stretchy as well. That might seem to contradict the idea of rigidity, but doesn't. For light to propagate over great distances, the ether had to be able to be distorted—this is in fact what defines a wave. In addition, light can be emitted in more than one distinct wavelength. All of this leads to the same conclusion: Ether must be very flexible.

The second characteristic of ether comes to us from Newton. Thanks to Newton, we've been able to accurately calculate planetary motion without accounting for ether. As a result, we must conclude that ether has no, or virtually no, effect on matter. Ether doesn't resist matter. In other words, matter can freely move about in ether without encountering the least bit of resistance.

OK, so, at this point, we have this stuff that is rigid as well as stretchy, but that offers no resistance to matter. Fine.

The third characteristic of ether is surprising to say the least: Ether is absolutely immobile. That's very interesting to scientists, because if ether is immobile, it represents a very special reference frame compared to all the Galilean reference frames that are equivalent to each other. Now that they finally have an absolute reference frame, the next question the scientists get all excited about is this: What is the speed of the earth relative to ether? One team attempted to answer this question. And failed—repeatedly—for a very good reason: Ether doesn't exist.

72. Michelson's Interferometer

Between 1881 and 1887, Albert Abraham Michelson, a German-American physicist, and Edward Morley, an American physicist, tried to calculate the speed of the earth as it moved through the ether. To do this, they developed a sophisticated apparatus called an interferometer, whose mechanism can be explained quite simply, though it's far more complicated in reality. Let's say you are on the roof of a train that's traveling at a constant speed, but you don't know the speed. And that's exactly what you want to figure out. If you have a gun that shoots tennis balls and you know the speed of the balls, then you can shoot balls in two opposite directions and measure the speed of the balls with respect to the ground. So, if you shoot a ball in the direction the train is traveling, it will go that much more quickly (known ball speed plus unknown train speed), whereas if you shoot one in the opposite direction, you'll see that it travels more slowly (known ball speed minus unknown train speed)—with respect to the ground. The rest is relatively simple mathematics and vector calculus—a high school student could do it.

In Michelson and Morley's experiment, it wasn't a matter of shooting tennis balls, but rather shooting light into mirrors in two perpendicular directions— while one shot would be enough in the train example, because you know which direction the train is traveling in, they needed two since no one knew which direction the earth was traveling in the ether. From there, the light was reflected back and, by measuring the phase difference in the reflected light, you can determine the speed and direction of the movement of the earth in the ether.

Galilean Relativity According to *MythBusters*

In one episode of *MythBusters*, the team carried out the following experiment: They used an air cannon to shoot a soccer ball at a speed of 50 mph out the back of a pickup truck traveling at 50 mph in a straight line. The scene is filmed from the ground, near the spot where the ball is fired. From this point of view, you see the ball fall straight down to the ground with absolutely zero horizontal speed, zero sideways motion with respect to the ground. The idea behind Michelson and Morley's experiment is the same.

Michelson was alone when the first experiment was carried out in 1881, but then he worked with Morley until 1887. They persisted for so long with the same experiment, because no one knew the way the earth moved in the ether: maybe it turned, maybe it stopped now and then, maybe it accelerated. . . . Therefore, they felt the need to take as many measurements as possible before coming to a conclusion. Especially since the results they got were unsatisfactory—so unsatisfactory, in fact, that they earned Michelson the Nobel Prize in physics in 1907, once the significance of the results was understood. This experiment should definitely be in the hall of fame of experiments whose failure contributed far more than if they had succeeded.

Because no matter how many times Michelson and Morley redid the experiment, they always got the same astonishing result: The light always traveled at the same speed, no matter what the direction of the apparatus and no matter when the experiment was performed. Or at least, the speed of light was always the same *within the margin of a precision error*—which is to say, as precisely as it could be measured—in different directions. This told M&M—yes, I call them that sometimes—for one thing, if the earth does actually move in the ether, its movement must be small.

The Austrian physicist and philosopher Ernst Mach was the first person to allege ether doesn't exist. We have to hand it to him: Ether seemed to create more problems than it solved, and at that stage it was just a mental construct, a necessary medium for propagating light waves. But apart from objecting to the existence of the ether, Mach hadn't achieved anything particularly impressive, and nobody really went along with his idea.

73. The Electrostatic Problem

Oliver Heaviside was a British physicist who in spite of being self-taught received the prestigious Faraday Medal in 1922. In 1889, he made an amazing and inexplicable discovery that would overturn Galilean relativity. Remember, Heaviside was no slouch in the field of electromagnetic phenomena. He's the one who reduced Maxwell's eight equations to the famous four equations that are still taught today (see pages 89–90).

Heaviside was studying electrostatic fields and noticed that when one of these fields was in motion, it contracted in the direction of its movement; it crumpled up on itself as it moved. Once it stopped moving, it resumed its initial shape. This was a *huge* problem. "Why?" you may ask, dear impatient reader. Quite simply because ever since Galileo, movement didn't make any difference—remember, standing still or moving at a steady pace, no difference—and that's the very foundation of relativity, allowing us to claim that there is no special inertial reference frame. But if I were moving with an electrostatic field, I would see that it contracts in one direction, so I would be able to tell, even without any other observation point, that I was moving—including during uniform rectilinear motion! And by looking at which way the field contracts, I could even figure out which way I was moving. This discovery rather shoots Galilean relativity in the foot—the ironic thing being that this revelation would end up being the foundation of the special theory of relativity—but I am getting ahead of myself.

The Irish physicist George Francis FitzGerald, a student and eventually professor at—yep—Trinity College, tried to provide an explanation of this phenomenon. And although his explanation came a bit out of nowhere and wasn't quite right, it still raised an interesting point. This is what FitzGerald thought: It's not just electrostatic fields that contract when they move—absolutely everything contracts, including matter. According to him, the effect was produced by the ether when a body passes through it. This hypothesis would effectively solve the problem, because if it were true, then when I move with the electrostatic field, I would also myself be compressed by the ether and, when observing the electrostatic field, it would no longer seem to be contracted, so I couldn't tell that I was in motion, and so Galilean relativity would be saved.

Actually, the idea behind FitzGerald's thinking was that matter is made up of molecules and intermolecular forces that are electric and therefore behave in a similar way to an electrostatic field. Remember that at that time, the electron hadn't been discovered yet, and there wasn't really any formal proof of the existence of the atom. So this is an excellent hunch on FitzGerald's part, even if his hypothesis for explaining the contraction of electrostatic fields was ultimately incorrect.

But there's another guy who came up with the same kind of hypothesis, and he laid it all out mathematically. His name: Lorentz.

74. Lorentz and Poincaré

Hendrik Antoon Lorentz, a Dutch physicist, developed the mathematical formulas for electrostatic fields contracting in length in the direction of motion, taking into account an absolutely immobile ether. To him, these contractions are completely "real," meaning he considers the moving electrostatic fields to be contracted in absolute terms.

The mathematical equations Lorentz formulated not only handled length contraction in the direction of motion but also going from one Galilean reference frame to another. These equations are called transformations. They were Lorentz's first transformations. Then something thrilling happened: Lorentz's mathematical equations, these famous transformations, allowed a relook at the still-unexplained results of Michelson and Morley's experiment. All of this was so promising!

These transformations were then further developed—you might say perfected—by another mathematician, the French physicist and philosopher Henri Poincaré. He completed them by adding two constraints: The speed of light is the same in all directions—this reinforces M&M's initial observations—and the speed of light is a speed that can't be exceeded.

His additions to Lorentz's initial transformations explain why many people, still today, consider Poincaré to be the father of relativity. Except that Einstein's genius wasn't part of these considerations on light. No. Because in fact, Poincaré thought there was an immobile ether and that length contractions are real. Actually, he doesn't say anything either way about the existence of the ether, but he saw it as a way to have, in mathematical terms, efficient equations. And note the very French chivalry of Henri Poincaré: He insisted that his improved transformations continue to be called the Lorentz transformations. Classy. And just at that moment, in walks Albert Einstein.

75. Einstein in 1905: The Third Article

We're still enjoying 1905, Einstein's *annus mirabilis*, when he published his third article, "On the Electrodynamics of Moving Bodies,"[181] which established the *special theory of relativity*. The article was published in September, just a few weeks after his earlier article on Brownian motion (see page 273). Its publication was naturally submitted to a jury of eminent peers, notably Max Planck, and it was relatively simple to read—except, perhaps, for the mathematical part. What you need to understand about Einstein is that he felt that the laws governing the universe must be elegant, the general conceptions of the universe must be elegant, and everything must fit together nicely.

Einstein is an aesthete, and ether made him sick! Ether introduced an ugly asymmetry: In mechanics, all frames of reference are equivalent, nothing is absolute; yet in electromagnetism, there is supposed to be this ether, which is absolute and immobile, but it must not interfere with mechanics. Einstein didn't like that at all. Plus, he could clearly see that apart from propagating light waves, the ether does nothing but cause problems—which he found all the more annoying because he had shown in his first article that light was made up of "grains" called photons (see page 51). While he didn't reproach his "almost" colleagues—remember that Albert was just an employee of the federal patents office in Bern—for having dreamed up such a thing, he thought that it wasn't needed, it was a bad idea, and it was time to forget about it. For good.

This meant he needed to show that an electromagnetic wave could propagate in a vacuum—and thus he laid the foundations of field theory—and then he had to answer all the questions that Lorentz and Poincaré were able to answer with the ether concerning length contractions and electrostatic fields in motion. He put in place a whole new theory of space and time. And they called it the *special theory of relativity*!

[181]   *Annalen der Physik*, vol. 322, no. 10 (September 26, 1905).

76. **A Problem with Clocks**

A little flashback, here. Helmuth Karl Bernhard, Count von Moltke, was a Prussian field marshal and a great strategist before becoming a member of parliament in the Reichstag in 1871. One of his achievements was writing several works on military strategy, and he said that one of the keys to a powerful Prussian army lay in being able to quickly mobilize troops by using the railroad system. But there was a problem. While the railroads served most of the kingdom of Prussia, each station had its own clock, which was set individually. When a train had to leave one station at 12:00 PM and arrive at another station at 4:00 PM, the clock at the destination might in fact say 3:56 or 4:07 when the train got there.

Von Moltke believed that synchronizing all the clocks in Prussia's train stations would be a magnificent way of demonstrating the unity of the kingdom of Prussia to the rest of the world. He therefore submitted this problem to the best minds, expecting nothing less than a perfect solution to this extremely tricky problem.

Many engineers and other technical folks became interested in the problem. Some tried a mechanical approach, which had no chance of success because of the large distances between the various stations. Others turned to solutions that would take advantage of their knowledge of electromagnetism. Some of these engineers happened to submit patent applications at the Bern federal patent office, where a level-three technical examiner by the name of Albert Einstein was on duty every day. Albert also happened to be the specialist for patents involving electromagnetism. And Einstein wrote up his third article. In it, he asked three questions in succession: the first one seemed ridiculous, the second one seemed stupid, and the third one proved his genius by making you realize just how subtle the first two questions were. Because he totally redefined what we call simultaneity and time.

First of all, Einstein posed the question of simultaneity: What does it mean if I say a train arrives at the station at 7 o'clock? His exact answer—which shows you how very readable the main text of the article is—is this:

If, for instance, I say, "That a train arrives here at 7 o'clock," I mean
something like this: "That the small hand of my watch pointing to
7 and the arrival of the train are simultaneous events."

See what I mean when I said "the first question seems ridiculous"?
Einstein was explaining to whomever was willing to read it—namely, Max
Planck—how to read time on a watch. He was laying the groundwork. Two
events are simultaneous if they happen at the same time and at the same
moment. In other words, there is simultaneity when there is contiguity.[182]

Hence Einstein's second question: What does it mean when I say a
train arrives at the station at 7 o'clock when I am sitting quietly at home?
Right away, you see it's not possible to give as precise an answer as for the
first question. When the small hand on my watch points exactly to seven,
is the train arriving at the station? Has it already arrived? Is it just about to
arrive? We can no longer talk about contiguity, of course, but what about
simultaneity? How can we be sure of the simultaneity of the two events?

And finally, this is the third question he asked: What does it mean
when I say that a train arrives at the station at 7 o'clock while I am traveling along on another train? The underlying question, following from
the previous one, is about simultaneity. While the second question was
about knowing whether we can talk about the simultaneity of two
events in different locations, this one asks about the simultaneity of two
events that are moving with respect to each other.

So those are the questions Einstein asked in his article, with a mind
to resolve the discrepancy between classical mechanics with its relative
reference frames—which had been working just fine for almost two
hundred fifty years—and electromagnetism, which was still quite
young—only about fifty years old—but very promising despite the ether,
which was absolute, immobile, and distinctly ugly.

As a result, Einstein put forward two postulates: The first one, as
already mentioned, was his rejection of the very idea of the ether. He
wanted to be able to suggest reference frames that aren't special with
respect to each other in electromagnetism as well. The second one, at

[182]   The state of being in contact with something.

the heart of a contradiction between classical mechanics and electromagnetism, is the idea that the speed of propagation of an electromagnetic wave is constant—an idea that directly follows from Maxwell's equations.

In response to the question "Constant, yes, but with respect to what?", Einstein posits that the speed of light, as of any other electromagnetic wave, is constant whatever the reference frame. Which is a completely crazy idea! And poses an enormous problem. Very fortunately, there is a simple way of explaining why this problem is enormous.

77. The Problem with Two Lights

Let's start out with the idea that the speed of light is constant. Next, imagine you have a gizmo in front of you. It has a button. The button is connected to two electrical wires; one goes off to your right and the other off to your left. The two wires are made of the same material and they are the same length. At the end of each wire is a light bulb; they are identical. You press the button. As you'd expect, both bulbs light up at the same time. And even if you imagine that they're very far away, the light from each bulb will take exactly the same amount of time to reach you—because you're right in the middle. There is simultaneity of the lighting of the two bulbs. During the same process, if an observer is standing right next to the bulb on the right, the light from that bulb will reach him faster than the light from the bulb on the left. That observer will see the bulb on the right light up before the bulb on the left does.

That's the problem caused by a finite, constant speed of light. If light propagated (travels) at an infinite speed—guess who thought that? His name begins with A—both bulbs would light up instantaneously for everyone. With a finite, constant speed of light, simultaneity of events becomes a matter of point of view, a matter of where you are. That's a problem, or seems to be anyway, because loss of simultaneity suggests that, depending on the point of view, the very principle of causality could be violated!

Loss of simultaneity is totally inconsistent with Newton's classical mechanics—that is, the rules we thought were *The Truth* about motion

Causality Principle

The causality principle is part of what one might consider the Ten Commandments of the Laws of Physics. Because it's a principle, it has never been proven, but neither has it ever been disproved by any experiment. The causality principle is the foundation all of science is built on.

The causality principle can be stated as follows—you'll see it's good common sense:

- No effect can occur chronologically before its cause.
- No effect can have a retroactive effect on its cause.

Let's get this straight: It's certainly possible for an effect to *reproduce* its cause—in that case, you're in a cycle of cause and effect—but it's never ever possible for an effect to influence its own cause ahead of time. With Einstein's theory, where the speed of light is constant regardless of the reference frame, we have to add another item: There cannot be less time between a cause and its effect than the time that would be needed for light to travel between the location of the cause and the location of the effect.

for more than two hundred years, the rules we just do not touch. Time is uniform and continuous, and space is fixed and absolute. All this drives the notion that simultaneity is preserved no matter where the observer is. To question simultaneity, someone better have a Hemi under the hood, to say the least! It's likely that if Einstein had been your typical, accepted member of academia, he might have been more timid about overturning so many established concepts—but he wasn't—so he did. And he did it in a big way!

Einstein started out with the idea that nothing was necessary or absolute. And from there, brick by brick, he built a new theory of space and time. The first brick he would lay was the only absolute in his framework: The speed of light in a vacuum is absolutely constant, regardless of the reference frame. To explain all the behaviors associated with light, as well as other discoveries—namely, the contraction of electrostatic fields—Einstein used the concepts of *length contraction, time dilation,* and

Einstein, the Scammer

Let's take another little detour to discuss something that's been going around ever since the internet became available to everyone everywhere. It's the idea that "they're lying to us." Some gullible internet users[183] are quick to believe the theories they find on the internet, even the craziest ones. These simple souls think they're in on some big secret—it doesn't bother them at all, it doesn't even occur to them, that literally *millions* are in on the "secret" too. But, whatever. Some of the secret theories out there are rather famous—for instance, the idea that man never walked on the moon because Stanley Kubrick was commissioned by NASA to create the images. Folks, this theory was started as an April Fool's joke.[184] And then there's the one about Einstein. According to this theory, Einstein wasn't particularly bright, and apparently, he abused his position in the patent office to steal the special theory of relativity from Poincaré—little does it matter that Poincaré never submitted a patent application to Bern. Someone started this theory because, just because, Einstein "didn't happen to be" in academia and was a complete unknown before he shared his ideas. And changed the course of the history of science. And revolutionized our understanding of the universe. Don't ya think, if he were a fraud, that someone—like maybe, Poincaré himself—would have called him on it at the Solvay Conferences[185] or when he received the Nobel Prize or during his many years at Princeton University? End of detour.

relativity of simultaneity. His theory was absolutely airtight and completely consistent with results from electromagnetism and classical mechanics. The introduction is over. Let's go!

---

[183]   Yeah . . . "some."

[184]   See *Opération Lune (Dark Side of the Moon)*, directed by William Karel (2004).

[185]   In 1911, Lorentz was the chairperson for the first Solvay Conference, which was attended by many physicists, including Marie Curie, Paul Langevin, and Ernest Rutherford, as well as Albert Einstein *and* Henri Poincaré.

78. **The Special Theory of Relativity**

The first thing that Einstein did was state: that space and time are inseparable by definition, that they are intrinsically related and cannot not be split up. Furthermore, space and time form a dynamic *continuum*, meaning they can change shape.

To understand how this works, we'll use a rather simplistic analogy; otherwise, it will be way too complicated. I'm sure you've already performed the following experiment. It consists of looking at the moon. I know that you know the moon is very large, but you can cover the whole moon with the tip of your finger. There's no inconsistency there, of course, and at no time did you ever think the moon was actually smaller than your fingertip. The simple explanation: Because of the distance between you and the moon, your visual angle of the moon is very small, making it appear much smaller than it really is (see page 37).

As I said, this is a very simple analogy. But it's essential for understanding what happens in a relativistic framework, because many of the observations are considered from an observer's point of view, which provides a unique perspective on what is observed. And this is a fundamental change from classical mechanics.

Next, there's another thing you should know about classical mechanics so you understand the problem it was for special relativity and the way to fix it: *velocity composition*—aka velocity addition—which we owe to Galileo. If you're standing on a train that's going 360 mph, even though you're standing nice and still on the train, you're actually moving with respect to the ground at a velocity of 360 mph—OK, folks, here's where we start using that nifty engineering short hand *wrt* for "with respect to." If you start walking on the train, let's say at 5 mph in the direction the train is going, your velocity wrt the ground is now 365 mph—or 355 mph if you walk the other way. This simplistic calculation is what they call the *velocity composition law* in Newtonian mechanics.

Finally, as we saw earlier in classical mechanics, assuming the reference frames are all valid, the following statements are all equivalent:

The train is moving forward at 360 mph wrt to the ground.

The ground is traveling beneath the train at 360 mph in the opposite direction.

The train is moving forward at 180 mph wrt the ground, which is traveling beneath the train at another 180 mph in the opposite direction.

Now let's switch things up a little, and instead of thinking about walking on the train, let's have you turn on a laser pointer and point it in the same direction as the train is moving. BTW, the speed of light in a vacuum is by definition 983,571,056 feet per second. To make our lives easier, and because we're not in a vacuum while we're on the train, we're going to say the speed of light is 186,000 miles per second—just FYI, if we use the same units for the train's speed, a train traveling at 360 mph is going 0.1 mile per second, or 528 feet every second.

To put the finishing touches on the procedure for our experiment, we're going to imagine a point on the ground near the track where we have an observer. We're going to call this point *point zero on the ground*. It's also totally permissible for us to define point zero as simultaneous for both the ground and the train because the train touching this point is an event that happens at the same place and time both from the train point of view and the ground point of view. Now, let's imagine you're at the back of the train. As the observer, you are going to start a stopwatch the exact instant the back of the train crosses point zero on the ground. We'll call this instant *time zero*. Then for this situation, we will say that, seen from the train, point zero on the ground coincides with the back of the train. And we'll call it *point zero of the train*.

The train moves forward on the track. When you pass point zero on the ground, you simultaneously start your stopwatch, turn on your laser, and point it forward—and so does the observer on the ground. As already mentioned, classical mechanics tells us that, from your point of view on the train, the light from the laser is traveling at 186,000 miles per second, but from the point of view of an observer on the ground, the train's speed gets added to the speed of the light from the laser. So, to the guy on the ground, the speed of the laser beam on the train is 186,000.1 miles per second, which is impossible according to Einstein. And, similarly, if the observer on the ground turns on a laser pointer, the observer on the train—that's you—will see that laser's light traveling at an ever-just-so-slightly-slower speed, 185,999.9 miles per second, also impossible according to Einstein.

This is where Einstein brings in *length contraction*. Einstein decides that when you observe something—a cat, a toaster, a satellite, a train, the ground—traveling very fast—at speeds approaching the speed of light—you see, *from where you are*, that something—a cat, a toaster, a satellite, a train, the ground—compressed in the direction it's traveling—which is what had been observed with electrostatic fields. You need to understand that contrary to what Lorentz and Poincaré suggested, there is no "actual" contraction of lengths. It's just a change in point of view. Like with the size of the moon, a simple change in point of view affects an observer's visual angle when he's looking at a body in motion. So to you—still on the train—regardless of the train's speed, a foot is always 1 foot long. When seen from outside, the something—a cat, a toaster, a satellite, the train you just missed, the ground—that's traveling very fast will appear contracted. Stay with me, this will be helpful very shortly . . . when the next train comes by.

Why did he need length contraction? Because during that fraction of a second it takes for the light to travel the whole length of the train, the distance it travels has to be the same, for you on board the train and for the observer on the ground. To simplify the writing and to get away from repeatedly saying "the observer on the ground," from now on I'm going to use the names of US presidents. Please don't read any political message into this; it's just because at least one US president happens to have cows on his ranch and cows happen to watch trains go by. During that fraction of a second, the train moved forward—a very tiny distance, but it still did move forward. So to George W. Bush the train appears a bit squished in length, and it just so happens this contraction—this squishing—adjusts things, perfectly compensating for the speed-of-light discrepancy described earlier! And so, at the end of this fraction of a second, the tip of the laser beam on the train is exactly one train length from Barack Obama. In this case, the speed of the laser beam on the train is, indeed, actually the same for you as for good ol' Ronald Reagan. This seems to be working nicely.

Just hold on a minute, though. It still has to work the other way around, from your point of view on the train. Bill Clinton's laser beam has to move at the same speed from his point of view and from yours. From his point of view, it's easy; his laser beam moves forward at 186,000 miles per second. But, for you it's a little more complicated. From your point of view, the ground is moving, and now the ground looks contracted, or shortened. So to you, not only is the ground moving backward but its length also looks squished—the railroad ties look skinnier and the spaces between them appear to be closer together. Consequently, after a fraction of a second, the light of the laser beam on the ground will have moved a lot less than a train length. From your point of view, we've only succeeded in increasing the difference between the laser beam on the ground and the laser beam on the train, as seen from the train.[186]

So, Einstein brought in another phenomenon, time dilation. According to him, space and time are connected dynamically, so a contraction of distance must be associated with time dilation, a way of stretching time. Once again, it's a question of perspective. If you're wearing a wristwatch and your vision is sharp enough for you to see the watch on George H. W. Bush's wrist, you'll see that the second hand on his watch moves a little slower than the second hand on your watch—in the same way, John F. Kennedy would see the second hand on your watch is moving more slowly than on his. Einstein said when you observe something traveling very fast—at speeds approaching the speed of light—from where you are, you see time there (on a clock on the thing traveling very fast) as passing more slowly. Here again, you must

---

[186] If you think this is complicated to read, imagine how complicated it is to write. And feel free to take your time, there's no hurry. This is important, it's how your universe works.

not think that time passes differently on the train than on the ground. To the observer that you are, a second always lasts one second.

For a moment, let's forget length contraction and talk about time dilation only. Instead of using a fraction of a second for our experiment, let's run our experiment until a laser beam has traveled one train length, either the beam from the laser on the train or the laser on the ground. This might seem like we're doing exactly the same experiment, but that's not the case at all. So let's go! The instant you pass point zero, you and Richard Nixon start your stopwatches and turn on your lasers. Seen from the ground, by the time the laser beam on the ground has traveled that legendary train length, the laser beam on the train, moving at exactly the same speed, hasn't arrived at the front of the train yet—because the front of the train has moved forward during that time—but it's not a problem because your stopwatch on the train, as seen from the ground, hasn't finished ticking the fraction of a second needed for your laser beam to reach the front of the train. But it will finish, the very instant your laser beam hits the front of the train. This seems to be working nicely.

Just hold on a minute, here. Again, this has to work the other way around, too, from your point of view. And it doesn't work. Because for one thing, the train moves forward, and for another thing, the two laser beams travel at the same speed, 186,000 miles per second, so a train length is traveled on the ground before a train length is traveled on the train—because the train is moving forward. And on top of that, Jimmy Carter's stopwatch runs more slowly, seen from the train, than your stopwatch runs, which means that his laser took even less than our aforementioned fraction of a second to travel a train length. Something is definitely not working here. And to understand what's not working, we have to take a step back from the things we're familiar with for a moment, so we can consider how at speeds approaching the speed of light things happen a bit differently.

Let's go back to our experiment and stop the action at time zero— we're not stopping the train so much as freezing time in our minds at that moment. The back of the train, where you are, coincides with point zero on the ground, where we find Dwight D. Eisenhower. And where is the front of the train? Here's where we come back to Einstein's second question. This is when we realize the extraordinary subtlety of his question. Is it possible to have simultaneity at the front of the train and at the back of the train at the same time? Here we go, and at the same time we can review our US presidents.

Due to the length contraction seen from the ground, Harry S. Truman sees the train as contracted—that is, shorter. From his point of view, the front of the train hasn't reached the line that marks one train length from point zero yet—if the train was stopped normally, the front would be exactly at that line. But from another angle, from your point of view on the train, the length of the ground is contracted—the distance is shorter—so the front of the train has already gone past the line at one train length from point zero. Look, Einstein or not, the front of the train can't be in two different places at the same time. There is a loss of simultaneity. Yet, if we mark where the front of the train is in these two situations, it's obvious that when it gets rolling again, its front will still pass the first mark and then the next.

How do we interpret this? Well—I must insist you stop checking your text messages while you're reading this, so you can focus for just a little while—Einstein said when a body moves in space, its time *shifts toward the past*. It sounds a little bizarre, but let's think about it for a minute. When the train was frozen at point zero, seen from the ground, the front of the train hadn't reached the line at one train length yet. If two passengers on the train, one in front and one in back, had synchronized their watches before the start of the experiment, because the front of the train hasn't passed the line at one train length from point zero, it means that, as seen from the ground, the wristwatch of the passenger in the front shows an earlier time than the wristwatch of the passenger in the back. That means—I emphasize—seen from the ground, when it is noon at the back of the train, it might only be 11:59:54 at the front of the train. So, when you observe something moving and this something is relatively long, then "its back is already more in the future than its front." I put quotes around that because if you spout this off to an expert on special relativity with no introduction, you'd probably be down on the ground so fast you might not even realize you'd just been punched.

Again, it's a matter of point of view. And actually, point of view allows us to explain length contraction, too. If you imagine that the front of the train, seen from the ground, is still in the past, to some extent, you see it as a little earlier than it is for the observer at the back of the train. Seen from the ground, the front of the train appears closer to the back of the train than it "truly" is—this "truly" is completely false, strictly speaking, because the train, strictly speaking, is as it is seen depending on where it's seen from. You might just say *it truly lies*.

Specifically, this means that if you put a stopwatch at the front of the train and synchronize it with one at the back of the train at the instant it leaves, when the train moves forward, the stopwatch at the front is just a hair behind the one in the back—needless to say, with regular trains traveling at normal speeds, it doesn't make a noticeable difference. We call this delay *relativity of simultaneity*; it explains the loss of simultaneity. We might think of it as a time shift.

Minkowski Space-Time

Though a four-dimensional mathematical space model was introduced by Poincaré in 1905, it was published by Hermann Minkowski, a German mathematician and physicist, two years after that. The paternity issue isn't a problem if you consider that Poincaré mainly considered it as a mathematical tool, while Minkowski used it to model space-time—and bonus—in a reality without ether. The Minkowski space-time, or more generally Minkowski space, is a four-dimensional mathematical space. In this space, various operations are possible—as in any mathematical space. Namely, certain transformations can be performed to shift one or more of the dimensions.

What's interesting about this model, at this point in the book, is that when we shift time in this mathematical space, since time is invisible to us, it affects the three other dimensions, which we can see. That's how we arrive at the inductive conclusions we get from our ongoing experiments.

Let's go back to our experiment and use length contraction, time dilation, and relativity of simultaneity. We're going to need more stopwatches and more observers this time. We'll have Madonna as another observer; we'll place her on the ground at exactly one train length away from dear old President Herbert Hoover. And we'll add Jeff Bridges[187] as another observer on the train. He'll be up in the front—you're still on duty at the back of the train. To help us get this, everyone has a wristwatch instead of a stopwatch. I plan to use times that are not at all to scale,[188] but they'll help with understanding what's going on.

We're still frozen at time zero. What have we got? As seen by President William McKinley, it is exactly 12:00:00 PM on his watch, the same time as on your watch, which he can see on your wrist at the back of the train—wow, good eyes. He sees that Jeff Bridges's wristwatch says 11:59:56 AM up front. Madonna's watch is perfectly synchronized with that of the illustrious president, precisely 12 PM. Unfreeze time. The train starts moving; immediately you and the former president light your lasers. When Grover Cleveland's laser beam reaches Madonna,[189] her watch shows exactly 12:00:04. To the only president in US history who ever served two nonconsecutive terms in office, the laser beam on the train hasn't reached the front of the train yet, because although the two laser beams move at exactly the same speed and the train is squashed in length, the train is also moving forward, and anyway, it's only 12:00:00 on Jeff Bridges's watch. When the laser beam on the train finally reaches the front of the train, Jeff's watch says exactly 12:00:04, just like Madonna's said when the other laser beam reached her. This seems to be working nicely.

But we still have to check if it works the other way around, as seen from the train. Let's freeze things at time zero again. From your observation post, at the back of the train, your watch shows 12:00:00 PM, just like Jeff Bridges's watch—he's in the same reference frame as you are; he doesn't move at all with respect to you. Because you are at point

---

187   Because he's cool!
188   The actual time it takes for light to travel a train length would be way too short.
189   This chapter is totally surreal, I admit that.

zero, you are simultaneous with President Benjamin Harrison, whose watch also says 12:00:00. Madonna, on the other hand, doesn't have the same time. But wait, remember, seen from the train, the ground runs the opposite direction from the train, seen from the ground. That's why in this case, Benjamin Harrison's watch is behind Madonna's watch, which is now ahead. It's already 12:00:02 on her watch. Unfreeze time. Lasers on. Since the ground and the laser beam are traveling in opposite directions, and the ground's length is contracted, the laser beam on the ground reaches its goal very quickly—it's the same speed as the other laser beam, but it traveled over a contracted—shorter—distance that's also moving backward—and so Madonna's watch shows 12:00:04. Jeff Bridges sets down his White Russian, looks at his watch, and notes that the laser beam reached him at 12:00:04. *Heureka!*[190]

It works both ways, seen from the ground and from the train!

We could redo the experiment with an additional observer in a truck moving wrt the ground at a slower speed than the train, in the same direction, or in the opposite direction, or even in a nonparallel direction. We could do this experiment a thousand different ways. The Lorentz transformations in Minkowski space will always prove that it works mathematically. The special theory of relativity lets us build models for any reference frame: All we have to do is use a constant speed of light (in a vacuum) and nonabsolute—aka flexible—space and time that are dynamically related.

Of course, we could whine and complain that Einstein must have been pretty full of himself to arbitrarily decide that the speed of light had to be constant, so he could go ahead and twist our familiar notions of time and space in all different directions, just so he could make everything "fall" into place. The reality was more way complicated. First of all, the idea that the speed of light is constant came from electromagnetic theories and from experimental results such as Michelson and Morley's experiment—you remember (see page 280)? M&M! I've been simplifying things as much as possible about what the special theory of relativity is. In his article, Einstein showed the reasons for what made it

---

[190] German for *eureka*!

possible for him to say what he said at every step in his logical process. In the dreaded words you've heard from every math teacher you've ever had: You need to show your work. And Einstein certainly did.

Even now, proof arrives daily to confirm Einstein's special relativity is a valid model for everything related to electromagnetism and quantum mechanics. And as an added bonus, special relativity is completely compatible with Newtonian mechanics, as long as the speeds don't approach the speed of light. When things get too close to the speed of light, classical mechanics won't work anymore. At speeds that are much lower than the speed of light, the effects of relativity of simultaneity are negligible; the equations are then tremendously simplified, so it works to use Newton's equations. It's beautiful, it's elegant, it's genius . . . in the true meaning of the word.

But there was a snag: Special relativity couldn't handle gravity. It had no way of managing reference frames that accelerated or that were spinning on an axis. Einstein wasn't happy about this limitation; he didn't like that gravity wasn't included. So he continued working to resolve this issue. It took him almost ten years to finalize what became, in 1915, the general theory of relativity.

GENERAL RELATIVITY

Through it, we rediscovered that we actually
understood nothing at all.

79. Newtonian Gravity

According to Isaac Newton, the gravity exerted by one massive body on another was a force that acts over an unlimited distance—although, as we learned earlier, gravity's strength rapidly diminishes with distance—and it acted instantaneously. And also according to Newton, if you placed the sun and the moon in a completely empty environment, a force would instantaneously be exerted by one on the other and vice versa. In 1905, Einstein took issue with this.

Yes, Einstein had just established his special theory of relativity; but, gravity as they knew it was a problem. Einstein said to himself, according to our understanding, if the sun disappeared instantaneously, the gravity between the sun and the earth would also disappear instantaneously. The earth would then cease to orbit, and lacking a reason to turn, it would take off in a straight line. Done. The problem is that this means the information about the loss of the attractive force of the sun would cross the distance from the sun to the earth—about 93 million miles—instantaneously; in other words, faster than light. Which is a big no-no according to special relativity. Einstein had already spent an insane amount of time on special relativity, to show that instantaneity or simultaneity had no meaning from an absolute standpoint. So he decided to attack the gravity problem. He was probably kind of excited about the idea of shooting down another Newtonian theory—he was hot on the trail.

When Newton had tried to explain what gravity was, he was content to "simply" express its action mathematically. You remember his

quote I included early in the book about what he thought about hypotheses (see page 41)? Well, guess what, I didn't give you the whole thing.
Here's exactly what Newton said:

> But hitherto I have not been able to discover the cause of those
> properties of gravity from phenomena, and I frame no hypotheses;
> for whatever is not deduced from the phenomena is to be called
> an hypothesis; and hypotheses, whether metaphysical or physical,
> whether of occult qualities or mechanical, have no place in exper-
> imental philosophy.[191]

Newton said the nature of gravity remained a hopeless mystery to
him, and he refused to imagine any cause whatsoever for gravity. And
just so you don't get the impression that I'm being stingy with famous
quotes, here is another interesting quote from Newton. In spite of what
he said in that last one, this shows just how brilliant he was:

> I do not know what I may appear to the world, but to myself I
> seem to have been only like a boy playing on the sea-shore, and
> diverting myself in now and then finding a smoother pebble or a
> prettier shell than ordinary, whilst the great ocean of truth lay all
> undiscovered before me.[192]

Gravity was a real head-scratcher because it was a force that wasn't
quite like the others. In addition, it was an attractive force that didn't
have an opposing force in nature—unique in mechanics, especially from
a distance. In electromagnetism, there are attractive forces, but there
are also repulsive forces. On top of that, gravity was a force with infinite
influence—there again, in electromagnetism, a positive charge creates
an infinite electric field with an intensity that diminishes with distance,
like with gravity. But unlike electromagnetism, gravity was a force that

---

[191]   Newton's *Principia*. Remember *philosophy* is the old word for "science."
[192]   David Brewster, *Memoirs of the Life, Writings, and Discoveries of Sir Isaac Newton*,
        vol. 2, chap. 27 (1855).

couldn't be constrained. It was impossible to build a cage to contain the force of gravity and prevent it from being felt outside the enclosure. Gravity seemed to be an irresistible force.

80. If a Roofer Fell Off a Roof

In May 1907, Einstein was still working at the patent office in Bern—yes, he had already published the four articles in 1905. He was starting to enjoy some recognition from the scientific community, but he continued to work in the patent office where he'd been promoted to a level-two examiner, and as he frequently did when he was thinking—daydreaming, his mind drifting peacefully, without even thinking about it—he looked out the window and saw a roofer working on the roof of a building down the street. Then a thought came to him that set in motion a whole new series of revolutionary thoughts. If the roofer fell from the roof, he wouldn't feel his own weight while he was falling. That seemed paradoxical, because it was his weight that was making him fall, but as soon as he yields to his weight, he doesn't feel it anymore. Einstein hadn't discovered anything new here. All of this was already expressed in Newton's equations, and to tell the truth, it had been understood by Galileo and his law of falling bodies (see pages 161 and 164).

It occurred to him: If a roofer falls off a roof and a few roof tiles go with him, the tiles would fall at the same speed as the roofer. From the roofer's point of view, the tiles "float" weightlessly around him. Einstein called the idea the "happiest idea of his life." It affected him physically—he felt a cold sweat and his heart started pounding. He knew he was onto something. He said:

> I was sitting in my chair in the Bern Federal Building. . . . I understood that when a person was in free fall, he wouldn't feel his own weight. It seized me. The idea left a huge impression on me. It drove me towards a new theory of gravity.[193]

---

[193]   Einstein's Kyoto address.

Einstein had just come to understand that being in free fall canceled, in a way, the effects of gravity, from the point of view of the person who is falling—and who is a Galilean reference frame—so, gravity isn't felt in free fall. Legend has it that Einstein once had a chance to talk to a roofer who'd fallen off a roof. He was quick to ask if the roofer had felt his own weight during the fall. The roofer replied, "Professor, I was just scared to death."

Einstein sought to understand the exact phenomenon that canceled the effects of gravity. And the first thing he found was the *equivalence principle*, which he revealed in a 1907 article titled "Relativitätsprinzip und die aus demselben gezogenen Folgerungen."[194]

81. Equivalence Principle

Imagine you're in a hut. It's on Earth, and it's all closed up, with no windows. The hut is large enough that you can conduct simple physics experiments, such as letting a disk slide down an incline, dropping a marble from a given height, and emptying a bottle drop by drop into a container directly below it. It's very similar to Galileo's experiment on the boat that helped him conclude that moving or not moving was simply a matter of point of view (see page 175). You do your experiments, and you confirm that your observations match the results of the calculations you made using Newton's equations. Very good.

Now put yourself in an identical hut, but in space. A place in space so remote and so far from any massive object that there are essentially no effects of gravity of any sort. Let's accelerate the hut upward—in the direction of the top of the hut, since obviously there's no up or down in space—and accelerate it enough so that the force from the acceleration is the same as the force of gravity on Earth. Let's go back and do our experiments under these new conditions. They will all work exactly the same as they did on Earth. Armed with this result, Einstein stated the *equivalence principle*, which we can paraphrase as follows: Locally, the effect of a *gravitational field* on a mechanical

[194]   Einstein's "On the Relativity Principle and the Conclusions Drawn from It".

experiment is identical to the effect of an acceleration of the observer's reference frame.

The equivalence principle says that the effect of gravity is identical to the effect of an acceleration; consequently, it's possible to use an acceleration to cancel out, or vice versa, produce the effect of gravity. So far, this seems to you to be sort of a beginning of something, but it's not exactly earth-shattering. That's because you're missing some information. In the preceding paragraph, there's mention of a gravitational field, but we haven't explained where it comes from. You're going to see that once we get that information—which, by the way, flows quite naturally from Newton's gravity—we'll be able to make the next giant step.

82. Geometrization of Gravity

Isn't "Geometrization of Gravity" a lovely title? It sure is a mouthful; it must be very technical, and complex, and hard to understand. Soon you'll see it's actually not—well almost. We learned, with Galileo, that falling has nothing to do with the mass of the falling body. Thus a lead ball and a cork ball fall at the same speed. We explained this phenomenon using the body's inertia—more mass, more inertia—and sometimes, we might completely confuse the concepts of mass and inertia. But we can still surmise: The weight of a body is proportional to its mass, and the weight of a body is the force that makes it fall. Given that the weight of a body increases with its mass, one might expect the speed of the fall to increase as well. However, the force of gravity, as expressed by Newton, also acts on the mass of the falling body. Mathematically, it's simple enough—don't worry.

The weight of a falling object is calculated using its mass (m_{obj}) multiplied by g, which is—we'll say it like this—the "value" of gravity on Earth (this is completely wrong, but it's good enough for now):

$$W = m_{obj} \cdot g$$

Gravity as expressed by Newton says this weight is proportional to each of the masses involved—m_{obj} for the falling object, m_{Earth} for the mass of the earth that attracts the object—and is inversely proportional to the square of the distance, D, between them:

$$W = G \: \frac{(m_{obj} \cdot m_{Earth})}{D^2}$$

G is a constant, called a *gravitational constant*, found experimentally by Newton. Looking at these two equations, we can see that g depends only on the following: the mass of the earth and the distance from the earth—in this case, the distance to the earth's center of gravity, which for the earth is its center:

$$g_e = G \: \frac{Earth}{d_{Earth}^2}$$

(The mass of the object m_{obj} isn't involved.)

Math time is over, I promise. The takeaway is this: The value of g at any point in space depends only on the mass of the earth and the distance between that point and Earth's center.[195]

I know you're thinking, gee, that's nice. So what does it do for me? And I'm telling you. The moon orbits the earth because of the gravitational force that the earth exerts on the moon. At any instant, the moon is a certain distance in space from the earth and also has a certain speed—with a trajectory that's tangent to the earth, which allows the moon to stay in orbit. If you take away the moon at any given moment, and replace it with Antoine Pinay—more on him later—and you give him the same speed the moon was traveling when you snatched it, Mr. Pinay would orbit the earth the same way—same distance, same

---

[195] Henceforth the value g will be known as *gravitational acceleration*, or acceleration due to gravity.

speed—as the moon did before he arrived. It's surprising—and I'm sure he'd be just as surprised as we are—but it's true.

This means that at any point in space, you can accurately calculate the acceleration due to gravity that would be experienced by a body in that spot, whether it's actually there or not. Which means you can map out the acceleration due to gravity at each point in space. The set of all these g values for all points in space is called a *gravitational field*. Just like a magnetic field determines the points in space that "carry" magnetism, a gravitational field determines the points in space that "carry" gravity. Once you put a massive

> ### Bond, Gold Bond
>
> After World War II, the French government's finances were in terrible shape. A sweeping policy of public investment begun in 1948 was struggling to show convincing results. So in 1952, breaking with that policy, Antoine Pinay, then minister of finance and economic affairs under President Vincent Auriol, got legal authorization from Parliament to make budgetary cuts of 110 billion francs. And to generate cash flow, he issued a particularly favorable national bond indexed to gold. As an extra added bonus—it was completely tax free! This bond was soon called the "Pinay Bond" by some, and the "Pinay Annuity" by others. Again, why am I telling you about this? Yes . . . why?

body somewhere in space, the gravity the body exerts on the space around it becomes a *property of space*. This is called the *geometric theory of gravity*.

From there—and this is where his real genius shines—Einstein tried to determine the way—why and how—gravity affects the space[196] around massive bodies.

---

[196] Note: Every time I said "space," I should really have been saying "space-time." But this was complicated enough without introducing time into the mix. From here on, sorry to say, we can't get away with that simplification anymore.

83. Non-Euclidean Space-Time

Einstein got stuck on the math—and if Einstein had a hard time with the math, you gotta know I'm not going to get too mixed up with it—and concluded fairly quickly that it would be impossible to develop the mathematical formulas for the effect of gravity in a Euclidean space.

Euclidean Space

A Euclidean space is a mathematical structure, a geometrical structure to be precise, in which all Euclidean geometry is valid. Unless you're a geometry buff—in which case you know what this is all about—all the geometry you have ever done in your life has been done within a Euclidean space. The world as we perceive it is a Euclidean space: two parallel lines never touch, the sum of the angles of a triangle always equals 180°, etc.

There are other ways to envision a space in mathematics. So we can call the surface of a sphere—for example, the earth's surface—a space. If you draw two meridians on this space, they are straight lines that are parallel . . . but they meet at the poles. If you start from a point on the equator and follow it a quarter way around the earth, then you immediately head north until you get to the North Pole, then once you reach the North Pole, you return to your starting point. Congratulations, you've just drawn an equilateral triangle that has three right angles! You can do geometry in such a space, you can determine its laws of geometry, etc. However, in that kind of space the laws of Euclidean geometry are no longer valid. This type of space is called a non-Euclidean space.

In a very general and very informal way, a twisted space is non-Euclidean.

But you see, Einstein wasn't a wizard with mathematics in non-Euclidean spaces. So, he got in touch with one of his old classmates, who was particularly fond of the subject. Einstein went to Zurich, Switzerland, so they could work on the new theory together. The name of the classmate: Marcel Grossmann.

84. Better Together in 1913: Coauthoring an Article

Einstein and Grossmann coauthored an article in 1913. This article was the precursor of what would later become the general theory of relativity. In this article, titled "Outline of a Generalized Theory of Relativity and Gravitation,"[197] they formulated a theory that presented a non-Euclidean space–time that could change shape and included gravity; they used a different approach, but it was still the same fundamental idea. This highly mathematical theory contained a particular equation that, like all equations, had an equal sign with some math on one side and some more math on the other side. One side of the equation described mathematically how the shape of space-time could change as a function of the density of the mass present, and the other side described how matter would move within this space-time. At this stage, we are very close to the 1915 version of the general theory of relativity.

Gravity, as presented in the article, was no longer a force as described by Newton, but a result of the change in the space-time's shape caused by the presence of mass. From then on, gravity was considered an intrinsic property of space-time. To Newton, space and time were absolute, completely independent dimensions that acted like a backdrop in a theater, totally unaffected by the play that was going on. However, to Einstein, space and time were dynamically interconnected—special relativity—and what's more, space and time had properties that varied as a function of the presence, or absence, of mass—general relativity.

This was truly earth-shattering. For the first time ever, space and time were no longer considered a picture frame within which we paint the laws of physics but instead were an integral part of the painting itself.

---

[197] *Entwurf einer verallgemeinerten Relativitätstheorie und einer Theorie der Gravitation*—casually called *Entwurf* ("Outline" in German) by those who are on a first-name basis with it.

Special vs. General

At first, the terms *special relativity* and *general relativity* can be a little confusing. I mean, special relativity is a universal theory about space and time, and general relativity is a theory about gravity. So why is special relativity called "special"? And why should general relativity, which seems to talk about something else, be the "general" version of relativity?

Now remember, Einstein's vision of the force of gravity was that it was merely the result of an acceleration, and he proved that with his equivalence principle. The special theory of relativity is called that because it's for the special case of inertial reference frames—that is, reference frames that don't accelerate. Once a reference frame accelerates, the equations of special relativity are no longer valid—suddenly, we get why gravity had never successfully been incorporated into this framework.[197]

Einstein's ingenious thought, as he pondered a roofer and some tiles falling from a roof, was that he could—or at least, he wanted to—think of the roofer as an inertial reference frame. He noticed that in his reference frame, as he was falling due to his own weight, (1) no force was being applied to him and, (2) to him, the tiles falling around him would appear to be standing still. His burning desire to include gravity at all costs in an inertial reference frame is what brought him smack up against the limitations of Euclidean space. So Einstein used non-Euclidean spaces to find a way to present gravity as consistent with the framework of special relativity, and thus Einstein, literally and mathematically, created a generalization of his first theory. And that's why it's called *general relativity.*

---

[198]   Does this hurt your head? Something we learn about in science class in our teens, something we take completely for granted, was not grasped by the greatest mind of the twentieth century until after years and years of pondering and monumental mathematical maneuvering.

85. Einstein in 1905: The Fourth Article

Admit it, you thought I'd forgotten about it, right? Unless it was you who forgot. . . . In 1905, Einstein wrote four incredible articles. The fourth article, "Does the Inertia of a Body Depend Upon Its Energy Content?," was published November 21, 1905.[199] In this article, Einstein did nothing less than develop the most famous equation in physics: $E = mc^2$. He showed that the energy in a particle at rest is directly proportional to its mass. In addition, the factor c, a constant equal to the speed of light in a vacuum—approximately 186,000 miles per second—is *squared*, which means that a particle, even with a very small mass, has a crapload of energy!

The huge difference between mass and energy means that with a small amount of mass, we can get a large quantity of energy that can be dissipated through radiation—that is, heat and light. That's where the principle of the atomic bomb comes from.

In this article, Einstein showed a number of things. Most notably, he showed that a particle that emits energy loses mass; this is consistent with the principle of conservation of energy. Indeed, when Einstein claimed a particle had intrinsic energy—not energy as it was then known, kinetic, potential, etc.—he had to make sure that if a particle emitted energy, it lost weight. Using his mass-energy equivalence, Einstein described the phenomenon flawlessly—a particle at rest that emits a quantity of energy, E, loses a mass, m, of E/c^2.

But one of the fundamental things that follows from this equivalence, the thing that's of particular interest to us in this chapter, is that gravity is more determined by the energy in a body present in space-time than by its mass, which is equivalent. And this is important: If the mass-energy equivalence shows mass and energy are synonyms for a body at rest, then we can look at energy instead of mass. We can then reason that the particles in a body—which aren't at rest and are, therefore, in motion and, therefore, have kinetic energy—will affect space-time and be affected by it, through gravity.

---

[199]   *Annalen der Physik*, vol. 322, no. 18 (November 21, 1905).

And light itself—composed of photons, considered to have zero mass and in constant motion at the speed of light—is affected by the presence of a massive body. Einstein was saying that light's trajectory could be altered by gravitational forces, even though light has no mass!

86. **And Mercury Proves It**

We already discussed, briefly, the problem of Mercury's orbit.[200] This orbit doesn't exactly behave as Kepler said it should. So, of course, after Kepler, there was Newton, and we have to keep in mind that other planets in the solar system interfere with Mercury's orbit—particularly oh-so-heavy Jupiter. But even taking other planets into account, Mercury's orbit, or more accurately the shifting movement of its nomadic perihelion, still posed a problem. Some astronomers tried in vain to show the problem was evidence of another planet, Vulcan. As I said, when I said it before, all in all, Mercury's perihelion problem was just a tiny snag in Newton's otherwise brilliant tapestry of planetary motion.

So, we just "made do."

This was where Einstein knew he'd have the chance to prove his theory was valid. He believed general relativity could explain the phenomenon that had stumped Newtonian mechanics for two centuries.

And in 1913, with the support of his friend, confidant, and BFF—and oh, right, Swiss physicist—Michele Besso, Einstein started in on the calculations. Using the general theory of relativity, which he'd continued to develop since the work with Grossmann, he planned to describe Mercury's orbit, including the disturbances to its perihelion. If you read the Einstein and Besso letters, you'll see a sequence of steps in their calculations.

Step 1: First, they made a massive mistake about the mass of the sun, so their results were completely wrong. FAIL!

---

[200] Over time the precession of the perihelion of Mercury makes Mercury's orbit trace a pattern like you might get with a Spirograph (see pages 103–105).

Step 2: This time, they were mistaken about the volume of the sun, so their results were completely wrong. FAIL!

Step 3: Here again, they made a mistake. They didn't account for Mercury's rotation, which isn't important in Newtonian mechanics, but in the relativistic mechanics of general relativity it certainly was significant. FAIL!—again.

They never did get it to work out.

Even though Einstein was going to use the calculation of Mercury's orbit to prove once and for all that his theory was valid, it's interesting to note that through it all, the failure to get convincing results[201] never caused him to doubt his theory. And he was right, because eventually, he did get exactly the right results using his theory—but I'm getting ahead of myself here.

87. The General Theory of Relativity

We're here! In 1915, Einstein wrote an article—which would be published in 1916—titled "The Foundation of the General Theory of Relativity."[202] In this article, he presented a full-up version of what he'd outlined in 1913. It offered a gravitational theory that could be verified by experiment. According to this theory, gravity was not a force as stated by Newton; rather, it was the manifestation of two facts: (1) Space-time changes shape in the presence of energy and (2) space-time controls how energy moves according to curving straight lines called geodesics— that is, they are mathematically straight, but to us they look like curves, just like Earth's meridians.

So this is what Einstein revealed: The earth doesn't revolve around the sun—WTF?! Actually, it goes in a straight line, in uniform rectilinear motion—WTF?! But actually, let's calm down, because really, it makes sense. If gravity isn't a force, then Earth isn't subjected to any force, meaning Earth is in inertial motion—really, you've got this. The genius insight

---

[201]   Did you see what I did there? A Pinay for your thoughts.
[202]   *Annalen der Physik* 4, vol. 354, no. 7 (1916).

was to point out that it's a straight line that's drawn in a space-time that's shaped by the massive presence of the sun. And yeah! Mind. Blown.

And Einstein's theory seemed totally validated by his accurate calculation of Mercury's orbit, especially because he had succeeded in determining the reason for the perihelion's precession. Logically, you're asking, wait, how did he do that if he repeatedly crashed and burned while he was working on it? Did he keep refining things until he got the answer? That's just about the same question David Hilbert asked—or it's the question he would have asked if he had known about all those years of errors.

What Is a Tensor?

With a thing as complicated as a tensor I'm sure not going to pretend I'm expert or anything. As an introduction, let's just say that those of you who know about tensors probably already have a migraine just from seeing the word. Those of you who don't will have a migraine by the end of this detour. Let's start with the understanding that what I'm going to say is absolutely approximate, and you can't go bragging to the first math teacher you come across that you've mastered tensors.

When you want to describe events or phenomena that occur in a three-dimensional space, depending on what you want to measure, you might get a number, a value. For instance, at every point in a room, you want to measure atmospheric pressure, or maybe temperature. You'd be measuring *scalar quantities*. Other than just collecting numbers, you may feel the need to measure something that's more than just a numeric value, something that has magnitude and direction—for example, measuring the influence of a magnetic field or a gravitational field at each point in a room. You'd be measuring *vector quantities*. There's a third option. Imagine an eraser in your hands, and you've decided to twist it. You're applying forces all over the eraser, which is what changes its shape. Now imagine a space in which this kind of torsional force is applied. Tensors are the generalization of this set of forces applied in each point in space.

It would be hard for me to simplify it any more than that. Sorry about that.

David Hilbert, one of the great mathematicians of the twentieth century, definitely at least as good as Henri Poincaré, had been interested in the question of Mercury's perihelion for a while. He was "this close" to producing a theory similar to general relativity in 1915—and had been practically drowning in his calculations. You gotta understand, these are incredibly, unbelievably, amazingly complex calculations. Einstein published one theory after another and solved the Mercury problem, and Hilbert sent him a letter. As you'd expect, Hilbert congratulated him on his successes, but he couldn't stop himself from asking how Einstein had found the answer to such a difficult problem in such a short time—and believe me, a difficult problem for Hilbert would be absolutely horrendous for most of us.

To our knowledge, Einstein never answered that letter. But to wrap our minds around how he did it, we have to introduce the most abstract concept in this book, the *tensor*.

In 1913, Grossmann gave Einstein a tensor he'd developed to do Einstein's calculations. Einstein studied the tensor a bit, but he wasn't happy with it. It lacked symmetry and elegance. It was, in a word, ugly. And Einstein refused to think that the universe behaved according to such an inelegant collection of rules. So he built his own tensor, elegant, symmetric, sophisticated, the most beautiful tensor ever, the George Clooney of all mathematical tools for generalizing vectors in n dimensions[203]—don't worry, I have no intention, whatsoever, of trying to explain it. Well, using this tensor made Einstein tear his hair out while working on that Mercury problem.

But by the time Einstein published his article in 1915, that tensor was history. Einstein decided he preferred the tensor Grossmann had initially proposed, henceforth called the energy-momentum tensor. And so using this tensor he rather quickly found the solution, based on the calculations he had been working on for more than two years—using a slide rule, no doubt; hand-held calculators weren't really a thing until the 1970s. In this case, he wasn't actually any faster than Hilbert. It was just that he'd already done most of the calculations, so all he had to do

[203] The tensor wasn't any good. It was beautiful, but it was wrong.

was plug[204] it into the new tensor—which may sound rather simple, but it's nothing of the sort.

In summary, general relativity showed that gravity was not a force. Rather, gravity is the expression of an interaction between energy density[205] and space-time. It is expressed by bending the space-time as a function of the energy density but also by distributing energy density through changes in space-time's shape. And for the hello-I'm-an-aesthete-my-equations-are-beautiful-please-love-me award, we have our winner! I can't resist the pleasure of writing this equation for you (I won't explain it):

$$R_{\mu\nu} - \frac{1}{2} Rg_{\mu\nu} + \Lambda g_{\mu\nu} = \frac{8\pi G}{c^4} T_{\mu\nu}$$

You're free to tell me that it doesn't make any sense, and I'll agree, but you must admit that, visually, it's beautifully simple considering everything it tells us, right?

88. Testing, Testing, 1-2-3

Albert Einstein definitely had to have some serious cojones to come out with these laws that were supposed to govern the motion of the universe in such a counterintuitive way. I mean, come on! Should we deny our own common sense and accept these harebrained ideas that nothing revolves around anything, that the planets move in straight lines? Worse than that, Einstein was saying that if you throw a rock in front of you, it moves uniformly in a straight line, but in a reference frame—in this case, the earth—that's accelerating. Anybody with eyes in their head can see that's a bunch of nonsense. Or maybe not. The general theory of relativity has survived rigorous testing; Mercury was but the first stone thrown at the theory during the ensuing uproar.

---

204   The complete phrase in engineering is "plug and chug."
205   It was the density of the mass at the beginning, but you remember: $E = mc^2$ . . .

For one thing, Einstein claimed that light could be bent by a very massive body. He was so confident, in fact, that he suggested a protocol to prove it happened. He suggested that during the next total solar eclipse a careful measurement be taken of the positions of stars that are visible in the area right around the sun. He claimed that these observations would give the wrong positions for the stars because the sun's mass would bend the light from the stars. On May 29, 1919, the British astrophysicist Sir Arthur Eddington attempted the experiment and found the exact differences in position that Einstein had predicted. Though at first glance this might not seem like a big deal, it was indeed a significant event, especially on a political level: Right after World War I, for an Englishman to confirm a German's revolutionary theory,[206] which completely changed our perception of the universe. . . . What a great moment, so full of hope. It was news that made newspaper headlines all over the world.[207]

Einstein predicted that a *gravitational lens* could also prove his theory was correct. The idea is as follows: Imagine a faraway galaxy perfectly aligned with a star and with the earth, and the star is between Earth and the galaxy. If the star is massive enough, it can bend the light that comes from the galaxy. It can bend it enough so the light can come around one side of the star, and around another side, in fact around all the sides—just like a regular optical lens can do, except this is a gravitational lens.

Einstein asked: What happens to space–time around a star if it is very massive or if its mass radically increases? The theory seemed to tell him that, beyond a certain threshold, space-time would be bent so much that everything that got too close would remain trapped, confined within close proximity to the star. Black holes were one of the obvious consequences of general relativity if it proved to be true. Of course, the biggest problem with black holes is finding them. If nothing escapes, not even light, how do we know they even exist? Through gravitational

---

[206] Einstein was German from 1919 to 1933 and also became Swiss.

[207] Except in France. In France, there had been a series of strikes, including newspaper strikes, between March and June 1919.

lensing, that's how. If the course of light is bent around an area of . . . apparently nothing, it's because this apparent nothing is very massive; it's a telltale sign of a black hole.

I'll shorten the long list of experiments that confirm the results predicted by general relativity (aka GR).[208] To this day, GR remains one of the most brilliant theories for explaining how our universe behaves. In 2011, thanks to an experiment that was as simple as it was expensive, *Gravity Probe B* confirmed the bending of space-time caused by the earth.

We talk a lot about bending space-time for visualizing the spatial changes, but don't forget about effects on time. If we didn't have a good understanding of general relativity, GPS wouldn't exist. Think about it: The clocks on the satellites in orbit definitely move a lot faster than objects on Earth's surface, like cars. General relativity is how they knew that they would have to sync the satellite clocks on a regular basis.

We are also beginning to anticipate, through general relativity, the idea that the universe isn't static—although Einstein remained convinced of a static universe. If it turned out that the universe is actually expanding,[209] the next question, perhaps more exciting than all the others before it, is this: Did our universe have a beginning?

89. The Big Problem with Relativity and the Rest

There is a big problem, which I shrewdly hid under the rug until just now. There are two branches of physics that are absolutely fascinating— one concerns galaxies, stars, and the universe, while the other concerns the infinitely small: atoms, particles, photons.

Currently, physicists who study the universe can use general relativity and all the relativistic mechanics to predict, observe, measure, characterize, and propose new hypotheses every day. Similarly, those who study particles can use crazy knowledge about quantum mechanics

---

[208] Yeah, that's what we call it once we're all chummy with general relativity. Yep, SR is for special relativity.

[209] It did, and it is.

and electromagnetism, all in the space and time framework provided by special relativity. But there's a problem . . . and it's not a little one.

At the scale of particles, general relativity isn't valid. Could it be incomplete? Could it just be wrong? In the same way, at the galactic scale, quantum mechanics ceases to mean anything, and even though gravity is still—by far—the weakest known interaction—the phenomenal amount of mass and energy involved make gravity dominant at the galactic scale. Maybe our knowledge of quantum mechanics is incomplete; maybe it's just wrong in spite of daily experimental evidence confirming its validity.

So as it stands, an astrophysicist must work within a different framework, with a different model, than a particle physicist. But what if we find ourselves faced with a gigantic mass confined in a tiny space? Like a black hole, and probably what the universe was like when it started to expand. In this kind of situation, we can neither ignore the quantum effects because of the ridiculously humongous scale, nor can we ignore the gravitational effects because of the tremendous amount of energy involved. But, ahem, general relativity and quantum mechanics are fundamentally and mathematically incompatible. To understand why, we need to understand what quantum mechanics is—but that's a whole other book. It would also require understanding what quantum mechanics is *not*—where the mass of physical bodies comes from; why the universe is expanding, faster and faster; where its mass is hiding; what solutions physicists are coming up with to resolve the compatibility issues between the two most valid theories known—to understand the epic search for the *theory of everything*.

Yes, another book is definitely needed to find the answers to these questions. So, until then, stay curious.

CHRONOLOGY OF SCIENTISTS

To be uber-classy, we've even gone so far as to include a chronology of some of the people mentioned in this book. No one will read it, but still, having one is very classy.

Pythagoras (ca. 570–495 BCE)
  Greek mathematician and philosopher
Parmenides of Elea (ca. 515–440 BCE)
  Greek philosopher
Anaxagoras of Clazomenae (ca. 500–428 BCE)
  Greek philosopher
Democritus (ca. 460–370 BCE)
  Greek philosopher
Plato (ca. 427–348 BCE)
  Greek philosopher
Gan De (4th century BCE)
  Chinese astronomer
Aristotle (384–322 BCE)
  Greek philosopher
Euclid (323–285 BCE)
  Greek mathematician and astronomer
Archimedes (287–212 BCE)
  Greek scholar, mathematician, and engineer
Eratosthenes (ca. 284–192 BCE)
  Greek astronomer, geographer, philosopher, mathematician, and director of the Library of Alexandria
Hero of Alexandria (1st century CE)
  Greek engineer, machine builder, and mathematician
Claude Ptolemy (100–170)
  Greek mathematician, astronomer, and geographer

John Philoponus of Alexandria (ca. 490– ca. 566)
  Greek grammarian and philosopher
Abu Ali al-Hasan ibn al-Hasan ibn al-Haytham (965–1039)
  Arab scholar
Pierre de Maricourt, "The Pilgrim" (13th century) French scholar
Nicolaus Copernicus (1473–1543)
  Polish astronomer
Ostilio Ricci (1540–1603)
  Italian mathematician and architect
William Gilbert (1544–1603)
  English doctor and physicist
Giordano Bruno (1548–1600)
  Italian philosopher
Francis Bacon (1561–1626)
  English philosopher
Galileo (1564–1642)
  Italian physicist, astronomer, mathematician, and geometrician
Paolo Antonio Foscarini (1565–1616)
  Italian theologian, mathematician, and astronomer
Hans Lippershey (1570–1619)
  Dutch optician
Johannes Kepler (1571–1630)
  German astronomer and mathematician
Christoph Scheiner (1575–1650)
  German Jesuit priest, astronomer, and mathematician

Willebrord Snell (1580–1626)
Dutch mathematician, astronomer, and geographer

Orazio Grassi (1583–1654)
Italian Jesuit mathematician, astronomer, and architect

Étienne de Clave (ca. 1587–ca. 1645)
French medical doctor

Antoine de Villon (1589—ca. 1647)
French philosophy professor

René Descartes (1596–1650)
French philosopher and mathematician

Pierre de Fermat (1601–1665)
French mathematician

Gilles Personne de Roberval (1602–1675)
French professor of philosophy and mathematics

Evangelista Torricelli (1608–1647)
Italian physicist and mathematician

Francesco Maria Grimaldi (1613–1663)
Italian philosopher, astronomer, optician, and philosopher of mathematics

Blaise Pascal (1623–1662)
French mathematician, physicist, and writer

Giovanni Domenico Cassini (Jean-Dominique Cassini) (1625–1712)
Italian astronomer, naturalized French

Christiaan Huygens (1629–1695)
Dutch mathematician and astronomer

John Locke (1632–1704)
English philosopher

Robert Hooke (1635–1703)
English scientist

James Gregory (1638–1675)
Scottish mathematician and optician

Nicolas de Malebranche (1638–1715)
French philosopher and theologian

Isaac Newton (1643–1727)
English philosopher, mathematician, physicist, and astronomer

Gottfried Wilhelm Leibniz (1646–1716)
German philosopher

John Flamsteed (1646–1719)
British astronomer

Benoit de Maillet (1656–1738)
French ambassador

Edmund Halley (1656–1742)
British astronomer

David Gregory (1659–1708)
Scottish mathematician and astronomer

William Wollaston (1659–1724)
English philosopher and moralist

Guillaume Amontons (1663–1705)
French physicist

Christian Wolff (1679–1754)
German lawyer, mathematician, and philosopher

George Berkeley (1685–1753)
Irish philosopher

James Bradley (1693–1762)
British astronomer

Leonhard Euler (1707–1783)
Swiss mathematician and physicist

Georges-Louis Leclerc, Comte de Buffon (1707–1788)
French writer and naturalist

Mikhail Vasilyevich Lomonosov (1711–1765)
Russian writer, poet, and scientist

David Hume (1711–1776)
British philosopher, historian, and economist

Nicolas-Louis de Lacaille (1713–1762)
French clergyman and astronomer

Pierre Charles Le Monnier (1715 –1799)
French astronomer

Emmanuel Kant (1724–1804)
German philosopher

Guillaume Le Gentil (Our Gentle Willy) (1725–1792)
French voyager and astronomer

Joseph Black (1728–1799)
Scottish chemist and physicist

Johann Daniel Titius (1729–1796)
German astronomer

Charles Messier (1730–1817)
French astronomer

Johan Carl Wilcke (1732–1796)
Swedish physicist

Joseph Jérôme Lefrançois de Lalande (1732–1807)
French astronomer

Charles-Augustin Coulomb (1736–1806)
French physicist and engineer

Luigi Galvani (1737–1798)
Italian physician and physicist

William Herschel (1738–1822)
English composer and astronomer

Anders Johan Lexell (1740–1784)
Swedish-Russian mathematician and astronomer

Antoine Laurent Lavoisier (1743–1794)
French chemist, physicist, and economist

Alessandro Volta (1745–1827)
Italian physicist

Giuseppe Piazzi (1746–1826)
Italian astronomer

Johann Elert Bode (1747–1826)
German astronomer

William Nicholson (1753–1815)
English chemist

Gian Domenico Romagnosi (1761–1835)
Italian physicist, lawyer, and philosopher

Franz Xaver von Zach (1754–1832)
German astronomer

Heinrich Olbers (1758–1840)
German physicist, physician, and astronomer

Karl Ludwig Harding (1765–1834)
German astronomer

John Dalton (1766–1844)
English chemist and physicist

Anthony Carlisle (1768–1840)
English surgeon and chemist

Georg Wilhelm Friedrich Hegel (1770–1831)
German philosopher

Thomas Young (1773–1829)
British physician, physicist, and philosopher

André-Marie Ampère (1775–1836)
French physicist, chemist, mathematician, and philosopher

Amedeo Avogadro (1776–1856)
Italian physicist, chemist, and mathematician

Hans Christian Ørsted (1777–1851)
Danish chemist and physicist

Carl Friedrich Gauss (1777–1855)
German physicist, mathematician, and astronomer

Johann Wolfgang Döbereiner (1780–1849)
German chemist

Alexis Bouvard (1787–1843)
French astronomer

Augustin-Jean Fresnel (1788–1827)
French physicist

Antoine César Becquerel (1788–1878)
French physicist

Michael Faraday (1791–1867)
British chemist and physicist

John Herschel (1792–1871)
English physicist, astronomer, and photographer

Karl Ernst von Baer (1792–1876)
Russian naturalist and professor of zoology and anatomy

Karl Ludwig Hencke (1793–1866)
German astronomer

Nicolas Léonard Sadi Carnot (1796–1832)
French physicist and engineer

Jean-Baptiste Dumas (1800–1884)
French chemist and politician

Wilhelm Eduard Weber (1804–1891)
German physicist

Urbain Jean Joseph Le Verrier
(1811–1877)
French mathematician and
astronomer

Robert Bunsen (1811–1899)
German physicist

Johann Gottfried Galle (1812–1910)
German astronomer

Claude Bernard (1813–1878)
French physiologist and physician

John Couch Adams (1819–1892)
English mathematician and
astronomer

Alexandre-Émile Béguyer de
Chancourtois (1820–1886)
French geologist and mineralogist

Edmond Becquerel (1820–1891)
French physicist

Rudolf Clausius (1822–1888)
German physicist

Paul Broca (1824–1880)
French physician

Gustav Kirchhoff (1824–1887)
German physicist

William Thomson, Lord Kelvin
(1824–1907)
English physicist

George Johnstone Stoney (1826–1911)
Irish physicist

William Odling (1829–1921)
English chemist

Lothar Meyer (1830–1895)
German chemist

James Clerk Maxwell (1831–1879)
English physicist

William Crookes (1832–1919)
English chemist and physicist

Dmitri Ivanovitch Mendeleev
(1834–1907)
Russian chemist

John Alexander Reina Newlands
(1837–1898)
British chemist

Ernst Mach (1838–1916)
Austrian physicist and philosopher

Edward Williams Morley (1838–1923)
American chemist and physicist

John William Strutt Rayleigh (1842–
1919)
English physicist

Ludwig Boltzmann (1844–1906)
Austrian physicist and philosopher

Thomas Edison (1847–1931)
American inventor and industrialist

Carl Wernicke (1848–1905)
German psychiatrist and neurologist

Oliver Heaviside (1850–1925)
English mathematician and physicist

Henry Le Chatelier (1850–1936)
French engineer and chemist

George Francis FitzGerald (1851–1901)
Irish physicist

Albert Abraham Michelson (1852–1931)
American physicist

Hendrik-Antoon Lorentz (1853–1928)
Dutch physicist

Henri Poincaré (1854–1912)
French physicist, mathematician,
and philosopher

Percival Lowell (1855–1916)
American astronomer

Joseph John Thomson (1856–1940)
English physicist

Nikola Tesla (1856–1943)
Serbian-born American inventor

Abbott Lawrence Lowell (1856–1943)
American lawyer and political ana-
lyst, brother of Percival Lowell

Heinrich Rudolf Hertz (1857–1894)
German physicist

Max Planck (1858–1947)
German physicist

David Hilbert (1862–1943)
German mathematician

Hermann Minkowski (1864–1909)
German mathematician

Wilhelm Wien (1864–1928)
German physicist

Walther Hermann Nernst (1864–1941)
German physicist and chemist

Hantaro Nagaoka (1865–1950)
Japanese physicist

Marie Curie (1867–1934)
French physicist and chemist

Jean Perrin (1870–1942)
French physicist, chemist, and
politician

Ernest Rutherford (1871–1937)
English physicist

Walter Bradford Cannon (1871–1945)
American physiologist

Bertrand Russell (1872–1970)
British philosopher and logician

Joseph Capgras (1873–1950)
French physician

Michele Besso (1873–1955)
Swiss engineer and physicist

Marcel Grossmann (1878–1936)
Hungarian mathematician

Albert Einstein (1879–1955)
Physicist, originally from Germany

Hans Geiger (1882–1945)
German physicist and chemist

Niels Bohr (1885–1962)
Danish physicist

Erwin Schrödinger (1887–1961)
Austrian physicist

Ernest Marsden (1889–1970)
Physicist from New Zealand

Walther Bothe (1891–1957)
German physicist, mathematician,
and chemist

James Chadwick (1891–1974)
English physicist

Irène Joliot-Curie (1897–1956)
French physicist and chemist

Frédéric Joliot (1900–1958)
French physicist and chemist

George Gamow (1904–1968)
Russian-born physicist

Clyde William Tombaugh (1906–1997)
American astronomer

Harrison Brown (1917–1986)
American chemist and geochemist

Richard Phillips Feynman (1918–1988)
American physicist

Clair Cameron ("Pat") Patterson
(1922–1995)
American geochemist

Giacomo Rizzolatti (1937–)
Italian biologist and physician

Boris Cyrulnik (1937–)
French psychiatrist and
psychoanalyst

Vilayanur Subramanian Ramachandran
(1951–)
Indian neuroscientist

**Dame Victoria ("Vicki") Geraldine
Bruce** (1953–)
British psychologist

Andrew ("Andy") Young
Contemporary British psychologist

Andy Ellis
Contemporary British psychologist

Dianna Cowern (1989–)
American-born science communicator. You should google her. Now.

ACKNOWLEDGMENTS

Because it's polite to say thank you.

To Mr. Bibas and the late Mr. Villain, who instilled in a goofy teenager (me) the desire to learn more—always.

To Alexandre Astier, who demonstrates every day that nothing is impossible for those willing to put forth the effort.

To Etienne Klein, whose humility allows him to share the spotlight with people like me.

To Richard Feynman—*because*!

To Gaëlle, without whom this book would never have existed—if you hate the book, it's all her fault.

To Jarod and Camille, to whom this book is dedicated.

ABOUT THE AUTHOR

Bruce Benamran launched *e-penser*, his YouTube channel devoted to explaining advanced science in simple terms, in August 2013. Since then, he has gained more than 1 million followers in his native France, and another 26,000 at his recently launched English-language channel. He holds a master's degree in computer science from the University of Strasbourg, and he lives in France.

Translator Stephanie Delozier Strobel is a mechanical engineer turned French translator. She lives in semi-rural Pennsylvania. She thanks her son, Tucker, for being her honorary millennial consultant, and thanks Ros Schwartz, Chris Durban, and Grant Hamilton for their encouragement in the craft of translation.